JOSEPH·JOSLIN
MEMORIAL·LIBRARY

BORN · DIED
JAN. 23, 1776 · JAN. 17, 1865
SETTLED IN WAITSFIELD
1798
THE GIFT OF HIS
GRANDSON
GEORGE·ALFRED JOSLIN
1913

THE
GENIUS
WITHIN

ALSO BY FRANK T. VERTOSICK, JR.

Why We Hurt
When the Air Hits Your Brain

THE GENIUS WITHIN

Discovering the Intelligence of Every Living Thing

Frank T. Vertosick, Jr.

HARCOURT, INC.

New York   San Diego   London

Requests for permission to make copies of any part of the work should be
mailed to the following address: Permissions Department, Harcourt, Inc.,
6277 Sea Harbor Drive, Orlando, Florida 32887-6777.

www.HarcourtBooks.com

Library of Congress Cataloging-in-Publication Data
Vertosick, Frank T.
The genius within/Frank T. Vertosick, Jr.
p.   cm.
Includes bibliographical references and index.
ISBN 0-15-100551-6
1. Intellect.   2. Adaptation (Biology)   3. Biological control systems.
I. Title.
QP398 .V47 2002
591.5'13—dc21   2001004456

Designed by Ivan Holmes
Text set in Monotype Janson

First edition
K J I H G F E D C B A

Printed in the United States of America

In memory of Robert H. Kelly, colleague and friend

CONTENTS

To the Reader xi

Introduction 3

PART I   THE SMALL

1   The Microbial Mind 21

2   The Immune Intellect 56

3   The Enzyme Computer 87

PART II   THE ABSTRACT

4   The Yang and the Um 135

5   Networks 177

6   The Selfish Neuron 205

PART III   THE LARGE

7   Wider than the Sky 231

8   Superorganisms 270

9   Magister Ludi 292

Addenda 324

Bibliography 336

Index 345

And this secret spake Life herself unto me:
"Behold," said she, "I am that which
Must ever surpass itself."

—Friedrich Nietzsche, *Thus Spake Zarathustra*, 1883

To the Reader

I recall seeing a simple little film many years ago about a man who lived among grizzly bears. In the closing credits, the following disclaimer appeared: "No animals were harmed in the making of this film." This seemed a rather curious assertion, given that the movie frequently showed big-clawed bears snatching salmon from a stream and eating them alive. Clearly, the writers didn't consider fish to be "animals," at least not animals meriting any ethical consideration.

This isn't surprising. While we claim to respect "life," it isn't life we respect at all. We respect *intelligence*. Our ethical systems are based solely on intelligence; the moral worth we assign a creature derives from its intellect alone. Bears are smart and fish are dumb, so the welfare of a bear comes before the welfare of a salmon. A man can be imprisoned for incinerating a single kitten, yet earn a nice living torching entire nests of young hornets. Insects and other supposedly imbecilic beasts garner little of our pity; brainless life forms—plants, bacteria, and viruses—don't enter our moral paradigms at all.

By equating brainpower with ethical worth, we place ourselves at the top of the heap—or so we believe. Ironically, although there is no accepted definition of intelligence and no foolproof way of measuring it, we are convinced that we are more intelligent than any other form of life, past or present. But what if we're wrong? What if we discovered that other creatures have comparable, albeit different, intellects? This would shake our religious and ethical frameworks to their foundations. Yet, surely such a supposition is absurd. Bacteria as smart as we? Thinking ants? But this book will dare to consider the "absurd" notion that even the lowliest forms of life have an immense capacity to learn, to compute, and to solve the most complex problems. By examining the way living machines reason, I will show that there can be no "unintelligent" life. From enzyme to ecosystem, bacteria to rainforest, all living things must think, and think well—or die.

This book will explore some difficult topics, but it's still written for the general reader. I will delve deeply into the mysteries of the biological thought process, but I'm not writing just for computer mavens or brain junkies. This work is intended for everyone interested in understanding our own place in the living world, as well as for those interested in life itself.

Two thousand years ago, the stoic philosopher Epictetus wrote: "When you close your doors and make darkness within, remember never to say you are alone for you are not alone; God is within, and your genius is within." Epictetus equates God with the "genius within," and I think he's right. Intelligence is the Divine spark that distinguishes the living from the nonliving, the ember that burns within all organisms, simple or complex, great or small.

This book represents the culmination of twenty years of personal exploration into the nature and meaning of intelligence. I am a surgeon by trade, not an expert in computer science or computational models, and so I write this from the perspective of someone who must deal with living systems firsthand, not just in theory. When it comes to understanding the true nature of life and intelligence, I submit that there are no experts yet, only eager amateurs like myself. So, while I may not get everything right, I at least try to be entertaining.

Enjoy. And think.

Frank T. Vertosick, Jr.

THE
GENIUS
WITHIN

INTRODUCTION
Brain Chauvinism and the Nature of Intelligence

"A brain is a very mediocre commodity. Why, every pusilanimous creature that crawls on this earth or slinks through slimy seas has a brain!"

—The Great Oz to the Scarecrow, *The Wizard of Oz,*
MGM Studios, 1939

I'm a rabid fan of the Firesign Theater, the four-man comedy troupe specializing in bizarre reenactments of old radio programs. Their eccentric recordings carry such schizophrenic titles as "We're All Bozos on This Bus" and "Don't Crush That Dwarf, Hand Me the Pliers." One of their older titles in particular— "Everything You Know Is Wrong"—still intrigues me to this day. Although the Firesign Theater intended their titles to be funny, I found this one vaguely unsettling.

I now know why I find the title so disturbing: it isn't just funny, it's also frighteningly close to the truth. Much of what we "know" is already wrong, or will be proven wrong at some point in the future. For example, nearly everything that a sixteenth-century scholar accepted as scientific gospel has long ago fallen by the wayside. Alchemy, phlogiston theory, Ptolemy's epicycles—these ideas now rest in peace in the great cemetery of wrongheaded science. How will our own knowledge be viewed five centuries hence?

Of course, no one can be wrong all the time. Nevertheless, those of us who seek truth in the physical world must always consider the possibility that we might be wrong. Just as Abraham was prepared to give his own son's life as an offering to God, so too must we be willing to lay our most cherished beliefs on the sacrificial altar of progress. This is a difficult task, even for the greatest of open minds. Einstein dismissed quantum mechanics because he could not accept a random universe ("God does not play dice with the world"). Linus Pauling, a pioneer in protein chemistry, thought that proteins, not nucleic acids, must be the carriers of the genetic code. (Although, as will be seen later, he wasn't far wrong in believing that enzymes, not genes, are the building blocks of cellular intelligence.) These brilliant men, who made their reputations by challenging the hardened beliefs of others, could still be supremely wrong.

Correct ideas must emerge, phoenixlike, from the spent ashes of wrong ideas. As sociologist Thomas Kuhn argued in his landmark treatise "The Structure of Scientific Revolutions," new concepts don't erode the established edifices of science like raindrops on stone, but bring them crashing to the ground with a single, resounding blow. To accomplish a paradigm implosion of this magnitude, new ideas must adhere so closely to the "everything you know is wrong" doctrine that they appear ridiculous when first proposed. The heliocentric solar system of Copernicus, Pasteur's microbes, Einstein's curved space-time—all of these ideas sounded absurd in their infancy. These theories challenged God, Euclid, even the evidence of our own eyes, yet they have survived the test of time.

As may now be obvious, I'm setting the stage for my own blow against the established order of things, a blow striking out at the conventional view of intelligence. In keeping with the Firesign

Theater's adage, I plan to take much of what we know about intelligence and call it wrong. In keeping with centuries of tradition, my ideas may, at times, sound ridiculous to the core; once my arguments have been made, however, I hope the reader will see past the veneer of absurdity and observe the deeper truths that lie below.

When I speak of intelligence, I mean the general ability to store past experiences and to use that acquired knowledge to solve future problems. I'm not limiting my discussion to human intelligence, which many consider synonymous with intelligence itself. Quite the contrary: I reject the notion that human intelligence is unique in the biological realm. Brains are good at solving a certain class of problems, but they hold no monopoly on problem solving in general. Science now labors under the misguided belief that intelligence is a property found only in hardwired conglomerates like brains and their electronic surrogates, computers. I call this misconception "brain chauvinism," and I will refute it by showing how all living things—even those entirely devoid of nervous systems—can (and must) use some form of reason to survive. In fact, I believe that intelligence and the living process are one and the same: to live, organisms (or communities of organisms) must absorb information, store it, process it, and develop future strategies based upon it. In other words, to be alive, one must think.

To see brain chauvinism in action, consider the following questions: Is a tree stupid? Does a population of bacteria reason? Can a cancer cell outwit a human oncologist bent on destroying it? At first glance, using words like *stupid, reason,* and *outwit* in the same sentences with creatures like trees, bacteria, and cancer cells seems little more than anthropomorphic nonsense. That's because the brain chauvinist within all of us knows that brainless creatures are bereft of any higher intellect—or any intellect at all, for that matter. But here again, everything we know is indeed wrong. In the ensuing chapters, I will demonstrate how bacteria, immune systems, ant hills, even single cells all display intelligence in their struggles to survive in a changing world.

In their own way, some living systems on this planet might even approach the vaunted intelligence of the human brain. And this isn't mere anthropomorphism. Bacterial populations don't mimic us; our brains mimic them, as shall be seen. If this now sounds a bit ridiculous, recall that I gave fair warning. It won't be the first time this book teeters into absurdity.

Brain chauvinism has three corollaries. First, because nerve cells evolved relatively recently, advanced intelligence must be a late adaptation of survival; second, because only multicellular animals have nervous systems, only multicellular animals can use the power of reason; and third, because it's the latest model to roll off the cerebral assembly line, the human brain represents the paragon of biological intelligence. All of these corollaries can be questioned, and I will deal with each as the book progresses.

But why attack brain chauvinism at all? Because it's wrong, and this wrong idea distorts not only our scientific perceptions but our spiritual ones as well. Brain chauvinism has a firm foothold in both biology and religion; creationists and evolutionists agree on one point, namely, that humankind is the most "advanced" form of life on the planet. We consider humans the final stone at the top of life's epic pyramid, the last chapter in biology's book. Just who built the pyramid or penned the book—God or Darwin—doesn't matter. All roads lead to us.

How do we justify our exalted place in the firmament? The answer seems obvious: we alone have the power of reason, or so we believe. Lower animals can show crude cunning, but only we, the brain chauvinists, can think. We ignore the complex behaviors of all other forms of life, attributing them to instinct or to simple feedback loops no more sophisticated than a furnace thermostat. We suffer from delusions of brain grandeur, awarding the power of reason to us and to us alone.

Human dominance in both the physical and metaphysical realms derives exclusively from our alleged intellectual superiority. What other reason for our supremacy can we conjure up? We're neither the fastest nor the strongest of creatures. We're not even the most prolific: in terms of biomass, ants are more plentiful, and they've been around much longer. Thus, our minds constitute our sole claim to biology's throne. In a very real sense, it is not humankind that occupies this throne but the power of reason itself. Should another creature (or machine) surpass us in cognitive ability, we would find ourselves instantly dispossessed. We embrace brain chauvinism out of necessity. It defines us and inflates our self-esteem as a species.

Our ethical systems assume the intellectual supremacy of human beings, assigning worth to other living things according to their IQs. We give ourselves souls but deny them to the apes; we readily kill dumb tunas but not smart dolphins. No one grieves for the dead house fly on the windowsill, yet everyone blubbers for "Old Yeller" when he takes a bullet. Some religions, like Hinduism, might assign worth to the fly, but I know of no religion banning the murder of dandelions or an HIV virus. All philosophical systems define some level of stupidity below which life has no ethical or moral value.

Yet, do we really know how smart any living thing is? Do we know, for example, that the dolphin is smarter than the tuna? Are we canning the wrong beast? In truth, we don't know. Some animal rights groups maintain that apes and dolphins should be assigned human rights. I'll steer clear of this debate. My point is simply this: in the final analysis, we don't fully understand the intellectual capacity of *any* living organism (or group of organisms). What's worse, we haven't even given the topic much consideration. We ruminate endlessly about our own intelligence and the intelligence of things like us, including primates and digital machines, while ignoring the massive intellect of the biosphere that surrounds us. Life has survived on this hostile orb for eons, righting itself after countless catastrophes. How stupid could it have been all this time? Life has given us the elegance of the genetic code and the infinite complexity of our own bodies, yet we hold chess-playing computers in greater awe. Scientists must either accept the genius of the living world or attribute the existence of life entirely to Divine Design. There is no middle ground. Mainstream evolutionists give us a false choice: the world was created either by an intelligent God or by unintelligent Chance. I would offer a third alternative: life was indeed created by intelligent design, but the design is entirely of earthly origin.

The brain chauvinist places the human nervous system at the center of the universe just as Ptolemy put the earth there, and for the same reason: because that's where we believe God wanted it. Both ideas are wrong. Do I mean to toss humankind completely from its pedestal? No; as I concede at the book's end, our brains do exhibit a certain uniqueness after all. But I also hope to promote the rest of our living, thinking planet to its rightful place.

Alongside us.

Years ago, researchers at a Welsh university received a grant of 65,000 pounds to determine if the old really are wiser than the young. Unfortunately, they spent the lion's share of this money defining the word *wisdom*. So, too, must I begin at the beginning and provide an operational definition of intelligence. Entire books have been written about this subject, and still no broad consensus has emerged as to what constitutes intelligent behavior. Luckily, the most contentious issues, like the precise measurement of human intelligence, are of little concern at present. For now, I'm interested only in the general nature of intelligence, a less controversial topic.

Prior to the twentieth century earthly intelligence could be found only in living things, and so the nature of reason is best explored, at least initially, within the context of biology. To survive, all living beings must respond to an incessant barrage of stimuli: good, bad, and neutral. Some stimuli are so potently bad they provoke an immediate, reflexive response. The abrupt withdrawal of my arm from fire is a spontaneous, non-analytical reaction; it requires no thought, or at least no complex thought. Unfortunately, most hazards can't be handled so simplistically. If I blindly leapt from every threat, I would soon exhaust myself. Moreover, some threats, such as a menacing animal, are better handled by walking slowly away. No creature can make it through life equipped solely with dumb reflexes. Reflexes alone do not constitute intelligence. Organisms must temper their reflexes with judgment, and that implies reason.

When reflex alone proves inadequate or counterproductive, living things resort to more subtle ways of dealing with environmental data. They begin by determining the predictive value of their experiences and storing those experiences for later application. In other words, organisms learn from experience and apply this knowledge to future challenges. Learning is central to all intelligent behavior.

But intelligence goes beyond rote memorization of past experience. Consider this example: a two-year-old boy is bitten by a Great Dane and, a month later, encounters a barking Chihuahua. He runs away. If the boy's intelligence consisted solely of exact memorization, he could react adversely only to another Great Dane. Instead, his brain does something much more remarkable: it generalizes the Great Dane into a broader class of short-haired, four-legged, sharp-toothed animals. Generalization is an indispensable part of learning, given that few new experiences mimic past experiences in every minute detail. Intelligent beings must be more than video recorders.

A child who burns herself on the fireplace will later cringe from a flaming match, even though the tiny flicker at the end of a tiny stick has no literal resemblance to the roaring blaze among a pile of thick logs. Intelligent beings extract key features from past experiences, identify patterns among those features, and later recognize the same patterns in novel situations. A burning match and burning log share a common pattern: heat, light, inconstant flickering, the blackening consumption of wood, and so on. Intelligent beings have the power to look beyond differences (the Chihuahua is smaller than a Great Dane) and see similarities (both have teeth and emit barking noises). The ability to see old patterns within new ones, even when the patterns differ significantly in other respects, is known as *pattern recognition* and is the cornerstone of biological reasoning. Indeed, we can now define biological intelligence as the ability to learn old patterns and extrapolate them to new situations.

Go to any fishing store and look at the lures. Some will be exact replicas of fish prey—grasshoppers, worms, and the like— while others look like mobiles hanging in the Museum of Modern Art. These latter lures simulate the patterns sought by hungry fish: flickering movement, shiny surfaces, critical color combinations, dangling hairs. Even the lowly fish responds to patterned information.

So far, this all seems rather straightforward, even obvious. Fish learn patterns exhibited by their prey and children learn patterns exhibited by dangerous dogs. The staunchest brain chauvinists will concede that fish have some intelligence because fish have brains, albeit tiny ones. We learn from observing patterns: patterns of our mother's speech, patterns of our golf pro's swings. Pattern recognition is the *sine qua non* of biological intelligence, and brains do it exceedingly well. The old television show "Name That Tune" was based on the human brain's uncanny ability to recall the pattern of an entire song after hearing only the first several notes. Yet, by equating intelligence with pattern recognition, I appear to be playing right into the hands of the brain chauvinists.

On the contrary. We can use this definition of intelligence to chip away at the foundation of brain chauvinism. Brains recognize patterns, that's true, but they haven't cornered the market on this talent—living things have been recognizing patterns since the earliest days of the primordial ooze. Pattern recognition is an abstract concept, in that there are no rules governing the nature of the patterns. So far, our examples (fish, dogs, little boys) use patterns made of up of macroscopic sensations tangible to nervous systems and computers: sights, sounds, smells, tactile sensations, and other information that can be converted into electrical signals. However, an intelligent machine can process any form of patterned information and use that information to guide its decisions. Trees decide when to sprout leaf buds based upon the patterns of light and dark, hot and cold that herald spring. Bacteria decide when to enter a spore phase based on patterns of moisture. Neither trees nor bacteria have brains.

Our immune system deals solely with three-dimensional molecular patterns, called antigens, located on the surfaces of bacteria and viruses. During our lifetimes, the immune system must learn and recall billions, perhaps trillions, of different molecular patterns. Our lives depend on its ability to make instant discriminations between friend and foe, not an easy task. And this isn't mere stupid reflex or inborn instinct, since tomorrow's microbes may be unlike anything the immune system has encountered before. The information flux through the immune system rivals that faced by any brain, and yet we still don't consider it truly intelligent; it's not even in the lowly league of brain-bearing invertebrates like worms or beetles. We, as macroscopic beings, have no intuitive feel for the molecular realm in which the immune system thinks, and so we fail to grasp its immense reasoning power. Therein lies a fundamental bias of brain chauvinism. To the chauvinist, intelligent beings deal only with macroscopic images, odors, sounds, words, numbers, and symbols. A bomb expert dismantling a terrorist's newest explosive is considered intelligent, while a bacterial colony dismantling one of our newest antibiotics is considered a pest and nothing more.

The immune system illustrates yet another bias in the chauvinist's paradigm: the centralization of intelligence. Because we identify "thinking" with central nervous systems, we can't fathom how decentralized and amorphous societies of cells, like the immune system (or a colony of bacteria), would ever be capable of rational thought. And we certainly can't imagine how a minuscule sac of enzymes—the cell—might think, even though dim-witted little enzyme sacs called ova manage to give rise to whole organisms like us. But if we look at what such societies accomplish on a daily basis, they must think, and think quite well. Our own brains are themselves nothing more than amorphous, decentral-

ized societies of cells. Although the brain and immune system encounter different patterns and seek different strategic goals, their
power of reason shares a common theoretical basis.

I'm intentionally avoiding the issue of consciousness, since it's
irrelevant to the arguments I'll advance in this book. Intelligence
and consciousness are not synonymous: intelligence is an observable quality; consciousness is not. A large urban zoo had a problem with a male orangutan escaping his pen and wandering the
zoo, terrorizing patrons as he searched for stray French fries. By
placing a video camera inside his pen, the staff discovered that the
simian Houdini had a paper clip he kept out of sight under his
upper lip. When he thought no one was looking, he removed the
paper clip, twisted it into a small loop, and picked the lock. He
then straightened out the clip and stowed it back under his lip. Is
this beast conscious? Who knows? I can't even tell if the man
seated next to me on the bus is conscious. But is he intelligent?
Most certainly.

Let's look at this episode in another way. Replace the living
orangutan with a robot ape made of microprocessors and servo
motors. Assuming the robot ape behaved in exactly the same way,
would we still consider it intelligent? Of course—any machine
capable of such deliberate planning and thought must be intelligent, regardless of its construction.

This was the same argument advanced by British mathematician and code-breaker Alan Turing in 1950. Turing, in an attempt
to define machine intelligence, proposed the "Imitation Game," in
which a human interrogator simultaneously converses, via tele-
type, with a machine and another human being. If, after a certain

length of time, the interrogator cannot consistently distinguish the human from the machine, the machine is deemed intelligent. Now known as the Turing test, this method depends solely on the observable consequences of intelligent behavior, while ignoring how that behavior arises. As Turing noted, we cannot "get inside" brains and machines to know if they are conscious, but we can directly observe their intelligent behavior. In the Turing paradigm, it doesn't matter what sort of contraption manifests intelligent behavior. The interrogator doesn't know or care if intelligent conversation originates inside a brain, a computer, or a bag of miniature marshmallows. If the conversation is intelligent, then the machine is intelligent. *Quod erat demonstrandum.*

I agree with Turing's black-box approach to the definition of intelligence. Unfortunately, however, Turing, as one of the planet's first true computer scientists, had a severe case of brain chauvinism. By its very nature, the Turing test identifies intelligence with language, particularly human language. Moreover, the human interrogator becomes the gold standard by which all other intelligent beings, living or machine, are to be judged. In the initial chapters of this book, I will conduct a series of "Turing tests" comparing our intelligence to the intelligence manifested by bacteria, the immune system, and cancer cells. However, I will modify the test by forcing us to communicate in the "language" of bacteria, immune cells, and cancer cells, rather than the other way around. When we have to reason on another creature's turf, we don't fare so well.

We must postulate a sort of Theory of Relativity for intelligence. In other words, how we define an intelligent machine depends on how we define the "frame of reference" that machine inhabits. What patterns of data must the machine sense, store, and recall? What class of problems must it solve, and how much time does it have to solve them? We can't evaluate a thinking machine unless we know its frame of reference. A molecular machine solving molecular problems may be as intelligent as a macroscopic machine solving macroscopic problems. Likewise, a slow machine dealing only with slow problems may be as intelligent as a fast machine faced with fast problems. As it turns out, living machines are as intelligent as they need to be in their unique frames of reference. The human brain isn't necessarily smarter than an ant's brain, or an immune system; it simply deals with a different kind of data and a different class of problems.

Back in the 1960s, Japan's exported products were generally inexpensive and of inferior quality. In the United States, the Japanese transistor radio was considered the epitome of shoddy workmanship. Many Americans began to think that postwar Japan was intellectually incapable of making decent manufactured goods. In reality, the quality of Japanese exports was dictated by American tariff policies that made it more economical for other countries to produce massive quantities of lousy merchandise. When the tariff laws were changed to make it more economical for Japan to export smaller numbers of big-ticket items like cars and expensive electronic goods, the quality of Japanese products soared. By the late 1970s, Japan's products set the standard for excellence, far surpassing the workmanship of comparable American goods. Thus, Japanese goods were as good as they needed to be in their economic frame of reference. Likewise, living things are as smart as they need to be in their unique realms.

How can a population of bacteria possibly "think"? And how can that population be favorably compared to a dedicated engine of thought like the human brain? Isn't a bacterial population nothing but a bunch of selfish cells obeying laws of natural selection at an accelerated pace? Yes, but our brains can be described in similar terms. As will be seen, *all intelligence is Darwinian.* When confronted with a problem, an intelligent machine chooses the "fittest" solution from a large set of competing alternatives. The main difference between the intelligence of evolution and the intelligence of the brain is the time scale. Compared to genetic evolution, the brain works significantly faster, of course, though not necessarily better. Moreover, the underlying mechanism—random noise coupled to some way of selecting the fittest variations—remains the same. As will be seen in the next chapter, there are ways to accelerate even a genetically driven evolutionary process and endow it with faster intelligence, without violating the basic tenets of Darwinian selection.

A creationist marvels at an orchid and asserts that only a Great Intelligence could have designed it. Quite true. Geological evolution looks intelligent because it is, and the human brain looks intelligent because it compresses a billion years of Darwinian trial and error into a few seconds. Mechanistically, brains, bacteria, and geological evolution all employ the same techniques to achieve their goals. Human thought is evolution temporally compressed a trillion-fold.

This book looks at intelligence as an *emergent property* of large groups. An emergent property is a property manifested by the whole group, even though that same property isn't apparent in any of the individuals comprising the group. For example, the temperature of a gas derives from the molecular motion of countless gas molecules, yet it makes no sense to speak of the temperature of one molecule. Temperature is an emergent property of a group of molecules. Similarly, human intelligence is an emergent property of a group of nerve cells. And immune intelligence is an emergent property of a group of immune cells, cellular intelligence an emergent property of a group of enzymes, and so on. To understand the relationship of intelligence to group behavior, we'll need to explore the science of *networks* in considerable detail. Networks are connected groups of individuals that can be used to store data and perform computation. I could summarize the main thesis of this book in one sentence: life is a network. The brain just happens to be one of the latest versions of a living network.

Networks form the basis of living intelligence at all scales of life, from cell to ecosystem. But I just said that all intelligence is Darwinian. How does network theory relate to Darwinism? Well, I can't explain everything in the Introduction. Let the journey begin. And let's hope that everything I know isn't wrong too.

In the first chapter, we'll take our first step and examine one way that natural selection can yield intelligent behavior, *à la* the Turing test. We turn now to the case of the Pharmaceutical Industry versus Bacterial Pneumonia...

PART I
THE SMALL

1

THE MICROBIAL MIND

"...the Martians—dead—slain by the putrefactive and disease bacteria...slain, after all man's devices had failed, by the humblest creatures that God, in his wisdom, has put upon this earth."

—H. G. Wells, *The War of the Worlds*

Tom was eighty-three years old. He ate sensibly, drank little alcohol, took his vitamins, and walked a local high school track for exercise. His sole vice was smoking: Tom had smoked a pack of cigarettes every day for over six decades. He had never given much thought to quitting; for most of his life, he felt fine.

But tobacco finally exacted a toll on his health in his later years. When he was young, Tom rarely got sick and never missed a day of work because of a cold or flu. Upper respiratory infections were now frequent visitors, each one harder on him than the last. When he arrived in our emergency room one chilly autumn evening, Tom could scarcely breathe. A recent bout of influenza had quickly progressed into bacterial pneumonia.

Influenza causes a raw inflammation of the trachea and bronchi, the air passages leading to our lungs. The inflamed surfaces ooze a viscous fluid, which is normally swept upward toward the mouth and nose by the incessant beating of microscopic hairs called cilia. After reaching the upper trachea, the fluid is coughed away. Viruses force us to hack and sneeze our way through our illness so they can spread to new hosts. By expectorating virus-laden fluids at other people, we complete their life cycle.

Cigarette smoke kills cilia, leaving denuded airways smooth as porcelain. Unhindered, a smoker's viral secretions slide downhill and collect in the tiny air sacs comprising the lungs. Pooled secretions become contaminated with airborne bacteria, and microbes quickly thrive in the warm, moist darkness. Although the immune system fights back valiantly (more on that in the next chapter), the defenses of elderly smokers can easily become overwhelmed by bacterial growth. So it was with Tom. His body could no longer combat the burgeoning population of bacteria infesting his diseased lungs and he sought the help of doctors, armed with modern antibiotics, to save him.

We, his doctors, would wage war on the bacterial cells now fulminating in Tom's body; it would be a war of modern medicine versus ancient, one-celled beasts. Our enemy, cunning and resilient, would not surrender without a fight. Nevertheless, we had the combined genius of the multibillion-dollar pharmaceutical industry in our camp.

For Tom, this would be a battle for a single life—his own. For the bacteria, it would be yet another skirmish in their eternal struggle against extinction. Could we, with our human brains, outfox the microbial mind, a vast, global intelligence flowing from the stunning plasticity of bacterial DNA? A better question: does a "microbial mind" exist at all? I say that it exists, for I have felt its presence.

We need to leave Tom for a moment and explore our battle against bacteria in some detail. Our chief weapons (besides our own immunity, of course) are antibiotics, substances that kill bacteria or halt their growth by crippling key components in their metabolic machinery; different classes of antibiotics attack different biochemical pathways. In this age of designer drugs, we often forget that the first known antibiotics were natural substances excreted by molds and other simple organisms. Ironically, some antibiotics come from the bacteria themselves. Biologists speculate that bacteria and molds use natural antibiotics to eliminate their competition and establish a foothold in overpopulated environments.

Most people are familiar with the legend of Sir Alexander Fleming. In 1928, Fleming, a microbiologist from St. Mary's Hospital in London, discovered that his staphylococcal cultures were being inhibited by mold contamination; he speculated that the mold, a common Penicillium species found in soil, was producing some chemical substance lethal to his staph. By sheer serendipity, Fleming had happened upon one of the miracles of modern medicine. He called his new substance *penicillin;* the discovery, although monumental, stirred little immediate interest.

Years later, a group of Oxford researchers led by Howard Florey teamed with an American pharmaceutical company to isolate and synthesize penicillin in the pure quantities needed for widespread clinical application. Their initial trials were startling: infections once thought to be incurable quickly abated after the administration of concentrated penicillin. The world rightly hailed penicillin as a godsend. The Oxford team didn't invent penicillin—Nature took care of that detail. However, they devised a way of delivering this natural substance into the human bloodstream at an unnaturally high purity and concentration.

The bacteria of the world were caught completely flat-footed by this development. To us, purified penicillin was a lifesaver, but to them, our enhanced antibiotics looked more like hydrogen bombs. True, bacteria had long encountered naturally occurring antibiotics, and a few species had even evolved defenses against them. Not surprisingly, antibiotic-producing bacteria proved particularly adept at detoxifying the poisons—spiders rarely get trapped in their own webs. Nevertheless, few bacterial species possessed a good defense against penicillin, and the defenses of those that did proved far too weak to handle the extraordinary doses we could now administer. And we didn't stop with penicillin. Twentieth-century medicine would soon unleash a slew of poisons into the environment, poisons more potent and diverse than anything previously encountered in the microbial world's billion-year history. Bacteria faced a veritable holocaust.

Microbial species that succumb to an antibiotic are said to be *sensitive*; those that survive its presence are called *resistant*. Modern studies conducted on species collected between 1914 and 1950 (the Murray collection) showed that bacteria from the "pre-antibiotic" era were universally sensitive to our purified antibiotic preparations, proving that pre-1950 microbes were ill equipped to handle the souped-up drugs that began appearing after World War II.

In the earliest days of commercial antibiotic use, our hopes ran high that we could win the battle against bacterial infections once and for all. We were brilliant humans, full of hubris, pitted against mindless simpletons. We had found the answer, and our foe was now defenseless. Even if bacteria could further evolve to counter our attacks, how quickly could they do so? Their pre-1950 antibiotic resistance, inadequate as it was, had taken millions of years to evolve. Now, in a matter of decades, they would be confronted with a bewildering array of concentrated doom. By the 1970s, chemists could modify antibiotic molecules almost at will, producing new drugs at a swift pace, and in the 1990s, we even enlisted the help of supercomputers to design new drugs. Now, not only do bacteria have to neutralize purified versions of natural substances, they have to deal with novel molecules of human design, molecules never before seen on earth. If they couldn't meet this challenge, they would soon be exterminated. The arms race was on. Could we produce new drugs faster than bacteria could evolve to defend themselves? Which would prove smarter, the pharmaceutical industry or the microbial mind?

To make matters worse for bacteria, we not only injected our drugs into sick humans but fed them to farm animals, sprayed them on crops, and dumped them by the boxcar load into our rivers, oceans, and ground water. The earth's bacteria, nearly a third of the planet's living substance, could easily have been annihilated. As brain chauvinists, we didn't give much thought to the welfare of single-celled creatures like bacteria, nor did we seriously consider what it meant to us if they all perished at our hand.

Imagine for a moment that Fleming had discovered a substance capable of killing every bird on the planet. We would treat this bird-killing drug like plutonium and transport it in a lead box under armed guard, not feed it to dairy cattle. Thankfully (since the vast majority of bacterial species are beneficial, not harmful, to life), bacteria rose to the challenge with a vengeance, and, I dare say, with no small amount of ingenuity. By contaminating the entire biosphere with our drugs, we incited the global network of bacteria to rebellion. Like some vast industrial community, they soon set to work deciphering our drugs and foiling our plans.

I'm reminded of a science fiction story I read as a boy, in which a scientist discovers a race of highly intelligent, though tiny, beings. He kidnaps them and establishes a colony in his laboratory, cultivating them like so many ants in a toy farm. One day, he's struck with a cruel inspiration: he places a gigantic weight over the colony and starts lowering it a fraction of an inch each day. He then provides his colony with bauxite—raw aluminum ore—and unlimited electric power. The beings, once content in their well-fed confinement, quickly realize they are soon to be crushed and commence making aluminum beams to shore up the weight. Ordinary aluminum proves too flimsy and the beams snap like twigs, but within weeks the beings invent a new aluminum alloy ten times stronger than anything known to humans. The scientist patents their discovery and grows rich. He proceeds to torment his pets in other ways, hoping to tap their ingenuity further. Though he can scarcely see the organisms who accomplish these feats, he knows they're intelligent because, in keeping with Turing, they produce tangibly intelligent outputs.

Fleming and Florey began lowering their great antibiotic weight onto the world's bacteria in 1940 and, in response, the bacteria immediately began throwing up their beams. Barely a year after penicillin's first commercial application, sporadic reports of bacterial resistance to the drug began to appear. Gradually, the beams hardened; by the mid-1950s, nearly every global species of bacteria merely laughed at the thought of penicillin, and the wonder drug was a wonder no more. Florey and Fleming garnered a Nobel Prize for a discovery that took them decades to accomplish, and some bacterial "simpletons" chewed up and spat out their miracle in less than half that time. And what prize did they attain for mocking our arrogance? Survival, the only prize that mattered to them.

If we conducted a Turing test using the language of pharmacology instead of words, who would look more intelligent to an impartial observer: Fleming, or his staph cultures?

◎

Before continuing, I need to provide a little technical background.

Living things are built from huge, carbon-based (organic) molecules known as macromolecules. In general, organic macromolecules are polymers formed by joining similar molecules into long chains. We now know that all genetic information is carried by polymers known as *nucleic acids;* included in this category is the now famous DNA (deoxyribonucleic acid), which consists of two polymer chains of deoxyribose (a sugar) twisted together in the "double helix" configuration first deduced by James Watson and Francis Crick nearly five decades ago. Bridging the two strands, like rungs on a ladder, sit a series of paired molecules known as the nucleotide bases. DNA uses just four bases: adenine, thymidine, guanine, and cytosine, abbreviated A, T, G, and C.

The pairing of bases is not random: A pairs only with T, and C only with G. Thus, if a strand contains the following base sequence, ATTGCATG, then the complementary strand must contain the mirror sequence, TAACGTAC. When DNA replicates, its twin strands untwist to form single strands; each strand then reconstitutes a new partner by attaching free bases in the appropriate sequence—an exposed A binds a new T, an exposed C binds a new G, and so on. In this fashion, genetic information is permanently "digitized" in the unique base sequence of a single DNA strand while the other strand merely goes along for the ride.

Another class of polymer macromolecules is the proteins, which are formed by linking together smaller sub-units, known as *amino acids;* all living proteins are built from just twenty different types of amino acids and may range in size from a few amino acids to long chains containing many hundreds. Unlike filamentous DNA strands, which tend to stay in a stringlike form, proteins typically fold themselves into complex globular shapes as the different amino groups adhere to one another. The intricate three-dimensional shape of a protein determines its unique biochemical functions. For example, the unique shape of hemoglobin—the main protein in red blood cells—contains a cleft just the right size for an iron atom. The iron, in turn, is used to carry oxygen. As will be seen in ensuing chapters, the ability of proteins to assume a variety of shapes is critical to the living process and to the generation of biological intelligence.

A protein's shape is determined entirely by its unique amino acid sequence. To understand why this is so, consider this analogy: assume we have a long metal chain and wish to twist part of it into a loop. We attach Velcro fasteners to two links and stick them together. The size of the loop will then be determined by which links have fasteners. If we place the fasteners twenty links apart, we'll create a loop twenty links in size. If, however, we place the fasteners only five links apart, the geometry of the chain may not permit us to make a loop at all. The stiffness of the links may prevent the fasteners from coming into close enough opposition to bind. If we distribute multiple fasteners along the chain and then tumble it in a clothes dryer, we will end up with a tangled three-dimensional shape determined entirely by the one-dimensional sequence of fasteners along the chain's length.

Certain amino acids act like Velcro fasteners, causing a protein chain to loop back upon itself. Thermal agitation at the molecular level acts like the clothes dryer, insuring that sticky amino groups will rapidly seek out and bind together. The position of the amino acid "fasteners" determines the protein's shape, but what determines the amino acid sequence of the protein chain is *the base sequence of its corresponding DNA gene.* (Although the terms *DNA* and *gene* are often used interchangeably, they are by no means synonymous. A gene is a sequence of bases that directly codes for the amino acid sequence of a single protein. Much of our DNA has nothing to do with coding for proteins and cannot be considered "genetic" per se.)

DNA encrypts an amino acid sequence using the *triplet code:* three sequential bases code for each amino acid in the protein's chain. For example, the triplet AGA codes for the amino acid serine. (Triplets are also called *codons.*) The triplet code is redundant, or *degenerate,* in that more than one codon can correspond to any given amino acid. In addition to AGA, three other codons code for serine. When a gene becomes activated, or "expressed," its base sequence is converted (translated) into the amino acid sequence of a real protein. Only a small fraction of a cell's genes are active at any one time; the rest lie dormant and unused. At the time of activation, the DNA gene unwinds and exposes its base sequence; the sequence is then copied (or, to use the correct term, *transcribed*) onto a piece of ribonucleic acid, also called RNA (a close cousin of DNA). In general, RNA is smaller than DNA and prefers to remain in a single-stranded form. These characteristics make it ideal for carting small chunks of genetic information around in a readable format. RNA is the workhorse of the genetic factory. If DNA is an encyclopedia, RNA is an office memo or work order.

After transcribing the gene sequence, this messenger RNA, or mRNA, leaves the DNA and journeys to the cell's cytoplasm, where it binds to tiny protein factories called *ribosomes.* A ribosome reads the mRNA, translates each codon in the base sequence into the corresponding amino acid, and then attaches that amino acid to the lengthening protein chain. (In addition to coding for amino acids, the triplet code contains "start" and "stop" codons that signal the ribosome when to begin and end a given protein chain.)

To summarize, protein synthesis requires:

1. *transcription* of genetic information from DNA to mRNA;
2. *translation* of mRNA into an amino acid chain (via the triplet code) at ribosomal protein factories; and, finally,
3. the *folding of the protein chain into its functional shape* as dictated by the linear sequence of amino acids.

This series of events—transcription, translation, protein synthesis—comprises the *central dogma* of molecular biology. In the central dogma, *a linear sequence of As, Ts, Cs, and Gs in a DNA molecule provides all the information needed to make a three-dimensional protein,* just as a linear sequence of 1s and 0s imprinted on a laser compact disc provides all the information needed to recreate the 3-D stereo sound of Beethoven's Ninth Symphony. The computer analogy is quite appropriate here, in that the genetic machinery functions as a vast informational engine; much (but not all) of life's knowledge now resides in nucleic acids. (This couldn't have been the case when life first arose on this planet, however, since life *predated* genes. In a latter chapter, we'll explore the unique relationship between genes and biological computation.)

The majority of our DNA doesn't code for proteins, but instead serves a regulatory function. Cellular regulatory proteins (such as those that bind steroid hormones) turn the various genes on and off by attaching to areas of "regulatory DNA" lying close to those genes. A large collection of genes, together with the intervening regulatory DNA, make up a single chromosome. Human cells have forty-six enormous chromosomes; bacteria have only one minute chromosome.

Amazingly, every living creature on earth, from the blind crab at the bottom of the ocean to the colorful rhododendron at the foot of Mount Everest, uses the same triplet code, and they all implement it in the same way. Human insulin was first manufactured for commercial use by inserting our insulin gene into bacteria and extracting the finished protein from bacterial broths. Bacteria transcribe and translate this mammalian gene into functional insulin protein just as our own cells do—at the genetic level, all life speaks the same language. From an informational point of view, this means all living things can participate in one, vast genetic community, provided they possess the means and the will to exchange DNA messages. Even "lowly" viruses adhere to the central dogma, but they don't waste their energy making their own mRNA, ribosomes, or proteins; they cleverly "borrow" the protein-making factories of other cells.

Viruses are little more than pieces of genetic material (either DNA or RNA) wrapped in protein. When a virus infects a cell, it sheds its protein wrapper and, once inside a cell, the naked viral chromosome acts like one of the host's chromosomes. Like an irksome neighbor who barges in and raids our refrigerator, viral DNA usurps the mRNA, ribosomes, and amino acids of its host and starts manufacturing new wrapper-proteins. After a slew of new wrappers are ready, the viral DNA replicates repeatedly and inserts a fresh viral chromosome into each wrapper; the new viruses then exit the cell in search of new hosts. The secretions that caused Tom's pneumonia came from bronchial cells plundered by influenza viruses.

Viruses infect bacteria as well as humans and play a key role within the workings of the microbial mind. Viral genes get sloppy during their replication, providing a pathway for genes to travel between organisms and between species. Viral nucleic acids often "integrate" into the chromosomes of their host, blending their genes imperceptibly with the host's and becoming a temporary part of the host's DNA. The viral genes then cleave themselves back out of the host DNA when the time comes for newly formed viruses to exit the cell. As viral DNA leaves the host chromosome, however, it may accidentally take a chunk of host DNA along with it. When the viral gene infects another host and again merges with its chromosomes, it may deposit the pilfered DNA permanently into that new host. (Many computer viruses work the same way, temporarily integrating into a computer's hard drive, then extracting some host data as it exits. This pilfered data may wind up on someone else's computer.)

We've learned to use viruses to shuttle DNA between cells to suit our needs. That's how we got the human insulin gene into bacteria: we tricked a virus into inserting the gene into a bacterial chromosome. Human gene therapy routinely employs viral vectors to carry new genes into ill humans who lack them. Viruses, in their sloppily promiscuous way, link all chromosomes on this earth into a single global chromosome of sorts. By infecting and reinfecting cells, viruses mix and match genes between different organisms, and even between different species, picking up stray genes here, depositing them there—shuffling the global genetic deck of cards, so to speak. If the stray gene proves detrimental to survival, the host dies and that genetic experiment ends, but if it proves beneficial, the host keeps it and passes it on to its heirs. If the new gene is supremely beneficial, it becomes standard operating equipment for that species. Many of our own genes were "field tested" in other species and transported to our chromosomes by viruses. We now know that viral gene transfer was a key factor in the rapid evolution of antibiotic resistance.

A brief word about mutations: a mutation is any deviation of DNA base sequence from the "correct" sequence. For example, a *point mutation* is a single base error—an A substituted for a C, for example—that occurs as a typo during DNA replication. Because the triplet code is degenerate, most point mutations are silent—they don't alter the final amino acid sequence of a protein. For example, given that both AGG and AGA code for serine, a mutation in the third position (from G to A or vice versa) will not change the protein; serine still appears in its correct position. The code is degenerate for precisely this reason, namely, to guard against error.

On the other hand, if AGC is changed to GGC because of an error in the first base location, then a new amino acid—proline—will appear in place of serine. This substitution constitutes a *missense mutation*. Sometimes even missense errors have little effect. The code is structured so that most single base mutations will replace an amino acid only with another chemically similar one. In this way, the code minimizes the impact of any one mutation on the final protein. Point mutations can still be dangerous. Sickle cell anemia, a deadly blood disease afflicting African-Americans, arises from a single point mutation in hemoglobin. Because of one missense error in the seventh link of an amino acid chain, sickle cell hemoglobin doesn't fold properly, distorting the red blood cell architecture. The malformed blood cells become wedged in tiny blood vessels, causing tissue damage, pain, and death. For want of a nail, a horseshoe was lost, etc.

Like any complex device, genetic machinery can go awry in any number of places. The ribosome may make punctuation errors, starting and stopping protein chains at the wrong places (*nonsense mutations*). Or, more disastrously, a mutation may "shift the reading frame" of codons by one or two bases, causing the whole gene to be parsed incorrectly. For example, the sequence ATT-TCG-GGC will yield one protein, but the same sequence parsed differently—A-TTT-CGG-GC—will yield another. If the ribo-some parses the triplets incorrectly, the resultant protein will be one-hundred percent wrong, a total monster. Like any language, the genetic code needs correct spacing and punctuation in order to convey an accurate message. The word *manage* and the phrase *man age* are quite different, even though they share the same se-quence of letters. A similarly misplaced space in a gene can wreck the whole protein.

Most mistakes in DNA replication, transcription, and/or translation are either neutral (they don't do anything) or lethal, and the likelihood that a faulty base substitution or ribosomal error will create a better protein is exceedingly remote. That's why bacteria often rely on something other than simple mutations when they need to evolve quickly.

When Tom arrived in our ER in 1995, we never considered starting him on garden-variety penicillin. For the vast majority of infections, penicillin no longer works, forcing us to rely on syn-thetic drugs intended to confound resistant bacteria. In the end, even these barely saved his life. It's tempting to explain this rapid emergence of antibiotic resistance in classically Darwinian terms. Dividing bacteria exhibit generational times as short as twenty minutes; consequently, a bacterial colony may grind through fifty generations in a single day. To put this into perspective, if we pri-mates had a similarly brief generational time, pea-brained homi-nids like Lucy would have evolved into modern Homo sapiens in about ten years.

Bacterial populations also contain gigantic numbers of indi-viduals, further increasing the likelihood that antibiotic-resistant mutants will arise spontaneously. With their massive numbers and rapid reproductive rate, bacteria become a great cauldron capable of cooking up genetic adaptations quickly. Or at least that's what mainstream Darwinian dogma asserts. If evolution is a crapshoot, bacteria come to the game with plenty of dice.

The standard textbook version of microbial natural selection reads something like this: bacterial populations are collections of discrete individuals, each mutating at a slow, uniform rate. Most mutations are either detrimental, ultimately killing the mutant cell, or neutral, meaning they don't affect the cell at all. The rare beneficial mutation gives the mutant cell a unique survival advantage over its peers. In a hostile environment, cells with beneficial mutations exploit this genetic advantage to become the dominant organism. The beneficial mutation soon becomes the mainstream version of the gene, and life continues. This all sounds very reasonable.

In this model, when a large bacterial colony is confronted with an environmental toxin, mutant bacteria arise that are, by purest chance, capable of neutralizing the toxin's lethal effect. Mutant antibiotic-resistant cells will survive in the presence of antibiotic, while their nonmutant comrades all perish. In a world (like ours) saturated with antibiotics, resistant mutants will become the sole species. This is precisely what happened in the case of strepto-mycin, an antibiotic discovered in 1947.

Streptomycin binds to ribosomes and wrecks the fidelity of protein translation, killing bacterial cells. (Streptomycin doesn't harm mammalian cells like ours, because our ribosomes differ slightly from bacterial ribosomes.) Within a year of streptomycin's first use for the treatment of tuberculosis, resistant strains of My-cobacterium tuberculosis (the causative bacteria) began to appear. The resistant M. tuberculosis cells carried a new point mutation in their ribosome genes that rendered their ribosomes immune to streptomycin's toxic effect.

But there's a problem with the simple natural selection model in the case of bacteria: one beneficial mutation—even two or three beneficial mutations—would rarely get the job done in time. The M. tuberculosis is unique in that it owns just a single copy of its ribosome genes. Thus, any mutation in those genes becomes readily apparent. By contrast, E. coli bacteria possess seven redundant copies of the ribosome genes. Thus, no single mutation in E. coli can render all of its ribosomes resistant to streptomycin.

To use the computer analogy again, E. coli's duplicate genes serve as backup files in its hard drive, shielding the bacteria from mutational miscues. Unfortunately, this backup system also shields E. coli from beneficial mutations like the one that confers streptomycin immunity. Recall that the genetic code is structured to make most mutations inconsequential. Given that mutations are rare, and beneficial mutations even rarer, the likelihood that all antibiotic resistance could have developed in only a few decades by mutation and selection alone approaches zero. Most methods of antibiotic resistance require the acquisition of entirely new genes (not mutant versions of old genes) and could not have arisen by mutation alone. Here we encounter the familiar criticism of mutation-driven natural selection. We understand how mutations can alter the beak of the finch, but not how they made a thing called a beak in the first place (much more on this in a later chapter). From where did these new "antibiotic resistance" genes originate?

The standard view of Darwinian evolution is a variation of the "one thousand chimps banging on typewriters for one thousand years can write *Hamlet*" theory. Evolution doesn't need brains in this scenario, only the clout of enormous time and overwhelming numbers of organisms. I agree with comedian Bob Newhart when he joked that the best chimps could do would read: "To be or not to be, that is the xgcjhgkeutotltlfg." Random mutation alone seems inadequate for the task at hand, whether it's writing *Hamlet* or inventing antibiotic resistance. Clearly, intelligence must be at work in both cases, but if we inject intelligence into something like the evolution of antibiotic resistance, we risk stepping onto the slippery slope of creationism by invoking an all-knowing Designer. After all, the evolution of bacterial resistance is a microcosm of evolution itself.

But there's no reason why evolving bacteria can't use cunning and brute force at the same time. Like professional gamblers at a Vegas blackjack table, bacteria know the game is random, but they also know a few tricks that can tilt the odds heavily in their favor. By examining the methods they used to "beat" the game of evolution and save themselves, we can get a glimpse of why life looks intelligently designed.

We put Tom in the intensive care unit and placed him on a mechanical ventilator to assist his labored breathing. Cultures of his sputum revealed two bacterial species growing in the depths of his lungs: staphylococcus aureus, the common staph bacteria, and E. coli, a species found in every animal colon. Forty years ago, penicillin would have devastated both species. Historically, the first patient ever treated with penicillin was a British policeman with a staph infection.

Bacterial cells have an outer casing, or cell wall, made from a mixed protein/sugar polymer called peptidoglycan, which cross-links into a stiff latticework, supplying the cell wall with its structural strength. Different species of bacteria have different types of walls—staph, for example, possesses a much thicker peptidoglycan cell wall than does E. coli—but they all use peptidoglycan as a type of microbial concrete. Several classes of antibiotic, including the penicillins, their cousins the cephalosporins, and vancomycin, prevent the cross-linking of the peptidoglycan lattice. In the presence of these antibiotics, dividing bacteria can still pour new concrete walls, but the concrete never sets up, never dries. Like houses built on liquid concrete, new bacteria literally fall apart.

Remember our Velcro fasteners on the protein chain? The cross-linking of the peptidoglycan chain requires one fastener, an amino acid called alanine, to bind to another alanine in a distant part of the chain. A molecule of penicillin mimics the alanine-alanine bond; penicillin acts like free-floating Velcro fasteners, tying up the protein's own fasteners and preventing the formation of vital alanine-alanine interconnections within peptidoglycan molecules. Penicillin can do other nasty things to a bacterium, too, but the inhibition of cell wall construction appears to be most important.

Bacteria generally resist the action of penicillin in two ways: (1) by denying penicillin access to the inner workings of the cell, where peptidoglycan synthesis occurs, or (2) by destroying penicillin molecules as soon as they enter the cell. Penicillin is a small molecule and is difficult to keep out of a cell, so most bacteria opt for the second method. There's also a third method: playing possum. Penicillin affects only cells undergoing division—it can't kill cells in a resting phase. Penicillin will not weaken fully formed cells, only prevent the formation of new cells. Bacteria may, in the presence of penicillin, simply cease dividing and play dead until the concentration of ambient penicillin goes down.

Penicillin molecules contain a structure known as a beta-lactam ring. Resistant bacteria learn to produce enzymes, known as lactamases, that shatter this ring, deactivating the drug. The family of lactamase enzymes is diverse, with each enzyme configured to attack one type of lactam ring. The broad family of lactamases couldn't have evolved via mutation alone in the short interval between 1941 (the first widespread use of penicillin) and 1942 (when lactamase resistance first became a problem). We now know that lactamases *didn't* evolve in that time frame—certain species of bacteria have owned lactamase genes for millions of years.

This seems to contradict my earlier statement that all pre-1950 bacteria were sensitive to penicillin. What I should have said was that all the bacteria we *knew* about—that were in our collections—were sensitive to penicillin. The vast majority of the earth's bacterial species have never been grown in a laboratory and, since the majority have no direct impact on us, we have never paid much attention to them. Among these unknown species are soil bacteria that learned to coexist with low levels of naturally occurring penicillin by evolving different lactamases over many millions of years. Bacteria that affect humans—the kinds of bacteria we know most about—never evolved any defense to antibiotics because our bodies never contained any, at least not until the last century.

Tom's staph species manufactured massive amounts of lactamase in 1995, even though the same staph species didn't possess this enzyme as recently as 1941. At some point during the past five decades, staphylococcus aureus acquired fully formed lactamase genes. They couldn't have generated this complex gene solely by mutational accident in that brief time, any more than chimps could write *Hamlet* in two years, so staph borrowed it from another source, namely, from one of the rarer, lactamase-producing species that evolved the enzyme long ago.

Gene exchange occurs in higher animals, like us, because we use sexual reproduction. But the common view of bacteria is that they are celibate loners, forced to make their solitary way in the world and adapting slowly via random mutation. In order to acquire the ability to make lactamase, staph not only had to acquire a new gene wholesale (and, in some cases, an entire chromosome of genes) from other bacteria, they had to acquire it from an entirely different species. If the earth were suddenly flooded, how easily could we swipe a set of gill genes from a species of fish?

It turns out that bacteria are anything but loners; on the contrary, there is so much exchange of genetic information among bacteria that the whole concept of "species" loses much of its meaning in the microbial realm. The transfer of genetic material between two bacteria can occur in one of three ways. First, as already has been seen, viruses can serve as the go-betweens, shifting genes between hosts; one species of virus, the bacteriophage, seems to have been intentionally engineered for this purpose. This viral transfer of genes is called *transduction*. Second, certain species, notably E. coli, use something akin to sexual mating in which the male bacterium extends a protein phallus to the female and injects her with some of his genetic material. Third, a bacterial cell may simply ingest genetic debris exuded into the local environment by dead comrades, a process known as *transformation*.

We now reach a crucial point in the story of the microbial mind, the part where we get our first glimpse of how bacterial evolution can transcend random mutation. Evolving bacteria still rely on mutation, but they employ more sophisticated techniques for countering new threats as well—techniques that accelerate bacterial evolution to such a degree that microbes assume a Turing-like aura of intelligence.

When legendary physicist Murray Gell-Mann was searching for mathematical symmetry among the fundamental particles of the universe—protons, electrons, neutrinos, and the like—he turned to an obscure branch of mathematics known as Lie unitary groups. Lie groups had been around a long time, but no one thought to apply them to elementary particle theory until Gell-Mann showed that one group, the SU(3), explained certain inter-relationships among particles. From SU(3) came the now famous "quark," and the rest is history.

The point of my physics diversion is this: Gell-Mann didn't invent group theory, he simply saw that it could be useful for his purpose. He needed a radical approach to a thorny problem, so he went to an existing body of mathematical knowledge (derived by other investigators for other reasons) and imported it into physics. After tweaking group theory a little, he found that it worked. We call this genius.

Science wouldn't advance very quickly if each scientist had to invent everything personally from the wheel up. Human accom-plishment depends upon the promiscuous exchange of informa-tion, an exchange that introduces nonlinearity into the system and permits human knowledge to grow exponentially instead of lin-early. We hear a lot about "nonlinearity" in computer science and biology these days. A system is nonlinear when the relationship between its input and output can't be described by a straight line. In mathematics, the behavior of a system becomes nonlinear when it feeds back into itself or when different parts of the system in-teract in a certain way. In general, nonlinearity introduces great complexity into the system. Consider an ecological system with three component populations: wolves, rabbits, and carrots. If the three components are isolated from one another, and provided with infinite food and space, the population growth of each can easily be predicted. But if we mix them together, rabbits will eat the carrots, wolves will eat the rabbits, and wolf and rabbit drop-pings will fertilize the soil for more carrots. Understanding the behavior of this nonlinear system will be considerably more difficult, if indeed the behavior can be completely understood.

The same complexity that makes nonlinear systems difficult to understand also endows them with an almost magical ability to get things done. Every interesting machine, including the human brain, relies on nonlinear behavior. Complex behaviors arising out of groups of interacting nonlinear entities are also called *emergent properties*. Such behaviors "emerge" mysteriously from the interplay of individuals in large groups.

Bacteria realized long ago that they could not survive if every bacterial species, armed with a paltry complement of only two thousand (or fewer) genes, had to invent its own solutions to every threat. They needed to introduce a greater degree of nonlinearity into their lives through communication and cooperation among themselves, including genetic exchange. When confronted with penicillin, staph went immediately to the genetic lending library comprised of all the world's bacterial DNA and looked up what it needed: a lactamase gene. From there, mutation and selection kicked in and customized the lactamase gene for staph's needs. One of the custom features was increased production of lactamase. Tom's staph was, by dry weight, almost 1 percent lactamase protein, enough to neutralize our biggest doses of penicillin. Like Gell-Mann, staph bacteria solved their crisis quickly by importing a body of information laboriously deduced by others for another purpose and customizing it for a new use. The ability to import large blocks of pre-evolved data across cellular boundaries hastened bacterial evolution immensely, allowing an incredible amount of adaptation in a short period of time. In humans, we call such accelerated adaptation to a new problem "reason."

The nonlinear interaction of bacteria goes beyond gene-swapping—bacteria also communicate using nongenetic signals.

Consider the case of the Vibrio fischeri. V. fischeri, a deep-ocean bacterium, produces a protein called lux, which, under the right conditions, emits light. But making this protein, and keeping it glowing, consumes a great deal of the bacterium's energy; consequently, V. fischeri don't bother making lux unless they are getting something really big in return. Specifically, they won't make it unless many other V. fischeri bacteria are around at the same time. It seems odd that a solitary bacterium could know (or care) how many other bacteria surround it; microbes don't strike us as social creatures. But V. fischeri both knows and cares. By signaling to one another using a hormone called lactone, the V. fischeri can estimate their own population density.

Each bacterial cell releases a small amount of lactone into its surroundings and, in turn, monitors the lactone level of its local environment. When V. fischeri density reaches a critical threshold, the ambient lactone level rises and triggers lux production. The entire population then glows in unison. The high densities of bacteria needed to create bioluminescence occur only within the glowing organs of deep-sea fish. There is a symbiotic relationship between the fish, which uses the glowing bacteria to see and hunt, and the bacteria, which feed off the spoils of that hunt. Only when inside a fish is it energetically profitable for V. fischeri to glow. In their free-living, low-density state, the bacteria have no need for light and won't waste their limited resources making lux protein.

This phenomenon, the simultaneous hormonal synchronization of behavior among many bacteria, is called *quorum sensing*. Each bacterium has a ballot; the population behaves according to how the majority of neighboring bacteria vote. Whether to glow or not to glow becomes a communal decision of the V. fischeri colony. Only the population as a whole knows when it is inside a fish; the individual cell can't know this. In the nonlinear vernacular,

bioluminescence is an emergent property of a V. fischeri community. I can't overstress these two points: (1) the intelligence of an
interacting, nonlinear community (what we'll later call a network)
greatly exceeds that of any single individual, and (2) life and intelligence are both, by their very nature, nonlinear processes arising from the behavior of large communities. We'll revisit both
points repeatedly in the coming chapters.

Hormonal signaling, such as quorum sensing, controls many
bacterial behaviors, including the willingness to exchange genetic
information. A cell's desire to seek out, acquire, and activate new
genes can be influenced by the needs of the community. Like bioluminescence, genetic exchange is an expensive business for bacteria. They may avoid it unless they have the backing of their peers.

We need to cover one last concept before finally returning to
Tom and resolving his struggle, and that's the concept of hypermutation under stress.

So far, we've viewed mutations as more bad than good, and the
reason for this is obvious: I'm unlikely to improve the quality
of the present manuscript with any misspellings, grammatical
mistakes, or typos. Similarly, a cell views mutations as ghastly
mistakes and thus avoids them—or at least corrects them. When
times are good, a cell prefers genetic stability.

DNA errors occur frequently, thanks to the mutagenic effect of
heat, natural oxides, ultraviolet radiation, and cosmic rays. As such,
all cells are equipped with an array of error-detecting and error-
correcting enzymes that serve as proofreaders, scanning genetic
material for mistakes and repairing damaged DNA. The normally
low rate of mutation derives as much from the efficiency of cellular proofreading as it does from the cleverness of the triplet code.

Evil as they may seem to us at first glance, mutations make the
difference between survival and death in periods of severe stress.

After all, change in the name of survival is what evolution is all about; *mutations and other forms of "noise" are a necessary resource for biological learning.* Coaches speak of "abandoning the game plan" when faced with certain defeat. A football team built around its running attack must start throwing the ball when behind fifty points to none. A tennis player used to playing the baseline, when losing badly, must charge the net. No sense sticking to a strategy that isn't working. Hypermutation—mutations occurring at a rapid rate—is the bacterial equivalent of totally abandoning the game plan. In this case, the game plan is written in their genes. Faced with imminent defeat, they increase their mutation rate and begin rewriting their DNA playbook in a hurry, hoping against hope for that quixotic touchdown, a mutation that saves them. In the genetic world, mutations are new ideas—not necessarily good ideas, but ideas all the same.

Even in times of plenty, bacterial populations harbor subpopulations with hypermutable genes, genes more prone to mutations during times of stress. These hypermutators are the Van Goghs of the microbial world—somewhat insane, but infinitely creative.

The hypermutator's increased rate of mutation may be due to an intrinsic fragility of its genes or to a flaw in the cell's DNA proofreading and repair mechanisms. Whatever the reason, the difference in mutation rate between a hypermutator and the average cell can be enormous. In starving E. coli populations, hypermutators arise with mutation rates up to one thousand times greater than normal. In the Darwinian paradigm, the rate of evolution depends upon the rate of mutation; accordingly, E. coli increase their rate of mutational evolution a thousand-fold during periods of starvation. In 1988, John Cairns and his colleagues postulated that stress-induced hypermutations occur only in genes associated with the appropriate stressor response. For example, during starvation, the mutation rate would increase only in genes responsible for food uptake and metabolism. For starving bacteria, reasoned Cairns, it makes no sense to increase the mutation rate in genes unrelated to food acquisition and metabolism.

Nevertheless, Cairns' theory of *targeted hypermutation* remains controversial, and the issue of whether stress-induced hypermutation affects only stressor genes or the entire bacterial chromosome has yet to be resolved. Some biologists fear that hypermutation targeted only at the appropriate regions of the bacterial chromosome endows bacteria with too much intelligence. Targeted hypermutation, if it exists, would put bacteria in total control of their own evolution, and this all sounds rather fishy to mainstream biologists. Brain chauvinism rears its ugly head again—bacteria can't be that smart. Or can they?

Targeted or not, accelerated mutation (and accelerated evolution) does occur in response to life-or-death situations, and we now have three sophisticated ways bacteria like staph and E. coli can influence the course of their own evolution:

1. acquisition of great chunks of genetic information via sexual exchange, viral-mediated transfer (transduction), or direct intake of genetic debris from the environment (transformation);
2. intercellular communication, which alters the behavior of large communities of bacteria and may regulate the exchange of genetic material as well; and
3. increasing the mutation rate up to a thousand-fold in response to stress.

By interfering with their own evolution, bacteria contribute additional nonlinearities to the process.

Of course, bacteria can still evolve in the stripped-down, no-frills way of using random mutation within isolated individuals. Many educated people consider this method—point mutations followed by natural selection of the fittest individuals—to be synonymous with evolution itself, but this process is entirely too slow to be the only way bacteria can adapt to sudden stress, especially bacteria that must make their living by invading hostile hosts. Random point mutation worked well for M. tuberculosis in its struggle against streptomycin, but that's the exception, not the rule.

Note that two of the three methods mentioned above—genetic exchange and intercellular signaling—require communication among different cells and even among different species of cells. As noted, this introduces nonlinearity, which then translates into the potential for more complex, emergent behavior. Even the third method—hypermutation—may be influenced by the social dynamics of a bacterial colony. The continuous presence of hypermutating subpopulations suggests that bacterial colonies display some form of social stratification with regard to their mutation rate, with different bacteria serving different evolutionary roles within a colony.

The old concept of bacterial colonies as random collections of unintelligent specks, each oblivious to the existence of its colleagues, must now give rise to the new view of bacterial colonies as organized, nonlinear societies comprised of tiny (but sophisticated) genetic machines connected by the constant interplay of hormonal signals and genetic information. Bacterial societies can address novel problems more intelligently than can any single bacterium, because they possess an informational architecture that allows large parallel populations of otherwise selfish individuals to work toward a common goal of survival. In an expanded time scale, we can consider every bacterial cell on earth as part of the same great community, the same network. Tom's staph may have gotten its lactamase gene from Japan, for instance. It couldn't have acquired it from Asia within hours, but it certainly could have within a few days, given international air travel. Rapid human transportation has further accelerated gene transfer among bacteria.

This makes the velocity of bacterial adaptation—in effect, the rate of bacterial thinking—dependent on jet planes and other human technology. True, but it's senseless to compartmentalize biological intelligence strictly according to species. Staph obtained its lactamase gene from another species of bacteria; it can exploit biped primates just as easily. How quickly we forget that we, too, are part of the natural world, and that our intelligence can't be easily divorced from the intelligence of the biosphere at large.

Bacteria form organized communities so they can respond to environmental threats more quickly than they could using sloth-like evolution in single cells. They still use Darwin's toolbox, but they use it more creatively. In bacteria, we see that the process of evolution has also evolved, but this should be no surprise—evolution seeks to improve every aspect of the living process, including itself. Through genetic exchange, hormonal communication, and hypermutation, bacteria have harnessed the fundamental principles of evolution and taken them to another level. For example, hypermutator clones of E. coli are bacteria that have literally evolved to be proficient at evolving. The conventional view of evolution is still correct, *but it represents only the first order process*—a purely linear form of genetic alteration. Over the eons, bacteria learned to add higher-order components to linear evolution, including hypermutation, gene exchange, and intercellular signaling.

The key to their success is communication, because communication allows bacteria to generate change exponentially, just as human communication generates knowledge exponentially. The modern view of genetic adaptation must take into account the co-operation among individuals, not just their competition. (Technically, competition is cooperation—negative cooperation—making competition a useful form of communication, too.) Communication enabled staph bacteria to acquire a million years of lactamase development in only one year, just as communication empowers me to install a computer operating system (which took decades for others to develop) in only minutes. Staph saw a solution to the penicillin holocaust by standing on the shoulders of giants, the shoulders of other lactamase-producing species. And we now have a name for the Darwinian adaptation exhibited by communicating networks: intelligence.

In a later chapter, I'll argue that our neuronal thinking process is no more than the latest version of accelerated evolution. A hint: nerve cells learned to mimic genetic variability using electrical signals in place of DNA base sequences. Another hint: nerve cells both compete and cooperate with one another.

Tom's bacteria, staph aureus and E. coli, are called pathogenic bacteria because they produce disease. Very few of the world's bacterial species make their living this way. Theirs is a hard life— it's much easier to feed off dead things. Living things fight back. Our immune systems are no pushovers; to breach our security systems and eat our lunch, pathogenic bacteria have to be as smart and nimble as jewel thieves. And eat our lunch is what they hope to do. We're food to them, plain and simple. I recall a comic strip depicting two emaciated vultures gazing sullenly upon a barren landscape. One vulture sneered to the other: "The hell with this scavenger business, I'm gonna kill something!" Pathogenic bacteria share the vulture's sentiments. They're tired of waiting for something to die, so they try killing things. Poor Tom was merely the human of the day.

As soon as they entered Tom's lungs, bacteria began deactivating his defenses. They also embarked on a war of attrition. As in human war, a bacterial conflagration often boils down to sheer numbers—our bodies can slay just so many bacteria per minute. If the bacteria can divide at a faster clip than we can consume them, *they* win.

When we use antibiotics, our aim isn't always killing bacteria; sometimes slowing their rate of reproduction gets the job done just as well. Recall that penicillin doesn't harm fully formed bacteria; it only prevents bacteria from dividing and making fresh bacteria. In war, to stand still equals certain defeat. If we can keep the population of bacteria static, our bodies can catch up and digest them, and then *we* win.

Laboratory tests confirmed that both the staph and E. coli in Tom's lungs possessed strong resistance to ordinary penicillins. The staph, however, was sensitive to oxacillin, a synthetic penicillin with a lactam ring engineered to survive staph's lactamase assault—score one for humanity. The E. coli was sensitive to gentamicin, a relative of our old friend streptomycin. We started intravenous infusions of both drugs, and Tom perked up briefly. Then his pneumonia relapsed. The staph had grown resistant to oxacillin by acquiring a new gene that denied oxacillin access to staph's protein factories. Likewise, the E. coli had acquired a new gene rendering it impervious to gentamicin as well as to several other drugs. The E. coli probably acquired a large resistance chromosome, known as a plasmid, by traveling to Tom's large intestine and copulating with bacteria there. Tom had been treated with gentamicin before and harbored resistant E. coli in his gut; those bacteria were only too happy to share their old tricks with new friends.

The staph could have acquired its resistance through hyper-mutation. Hypermutating clones may have stumbled upon a way to exclude oxacillin, then died and shed their genetic knowledge into the pus filling Tom's lungs. Other bacteria, rummaging through the wreckage like collectors at a garage sale, scooped it up and used it. Alternatively, a bacteriophage could easily have airlifted the new gene into Tom's lungs from a patient in another part of the hospital.

Meanwhile, a mortal conflict raged inside Tom's inflamed air sacs. Staph talked to staph, E. coli talked to E. coli, E. coli even talked to staph. Plots were hatched, genes flew, hormones were secreted, mutations rates rose and fell. Bacteria from the atmo-sphere, other patients, a doctor's hands, the colon, the blood-stream, and even Tom's skin all eagerly joined the fray, pitching in free DNA advice via bacteriophage transduction and DNA trans-formation. Variations in DNA sequences and different protein configurations were begged, borrowed, stolen, invented, and field-tested in a matter of hours. Hundreds of proteins, hormones, genes, and DNA regulatory mechanisms came into play, a nonlin-ear symphony of information beyond the ability of our best su-percomputers to simulate. Within every droplet of Tom's sputum, a myriad of different cell types fought or played a supporting role: bacteria, viruses, lung cells, and white cells of every stripe.

Tom's immune system quickly unloaded a barrage of pre-formed defense molecules, then set to work crafting unique anti-bodies to attack molecules on bacterial cell walls. In turn, the bacteria modified their cell walls and intensified their assault. They used ancient rules of engagement in their fight, but they were not actors blindly following a script. These were thinking beings, rapidly reacting to new situations with new strategies. The rules of chess are ancient, too, but any given game can be brilliantly and originally played. The human brain operates according to ancient rules, but that hardly nullifies the genius of Mozart. We administered other drugs, including a few more synthetic penicillins, and the crafty bacteria quickly deciphered these, too. Remember the boy running from the Chihuahua? Bacteria use pattern recognition in the same way; they sense the similarity of our doctored versions of penicillin and alter their enzymes accordingly. Penicillin, oxacillin, methicillin, ampicillin—they all look like dogs to bacteria.

The longer Tom lingered in the hospital, the greater access his infecting organisms had to the handiwork of other microbes. Hospitals teem with resistant bacteria. Finally, we tried ciprofloxacin, a newer agent, and this proved too much for Tom's pathogens. Ciprofloxacin halted both bacterial species long enough for one final charge by Tom's white cells. The infection ceased. Faced with a strengthening war on two fronts, the bacteria capitulated. Tom would live to fight another day. The bacteria retreated to lick their wounds and begin their work on a global solution to this latest threat. It wouldn't be long before they solved this drug as well. Another Tom, on another day, would feel their wrath.

If I speak with admiration for these creatures, it's because I've won and lost many battles against them and I *do* admire them. We have long underestimated bacteria, and they are now making us pay for our arrogance. Whatever their form, we must honor the intelligence of living things. To do otherwise is both disrespectful and dangerous. Unlike nonclinical biologists like Stephen Jay Gould or Richard Dawkins, physicians like myself enter the competitive arena and do battle with supposedly unintelligent beasts like bacteria and cancer cells. Darwin observed the creatures of the world with a keen eye, but he never fought them one on one. For those of us who stare into the shining eyes of the world's predators, we know how cunning they are at what they do. The bacterial resistance that almost killed my patient was not the handiwork of a "blind watchmaker" (Dawkins' phrase), nor was lactamase the product of a "selfish gene" (another Dawkins phrase) seeking only to preserve itself. The genes, enzymes, and bacteria in this saga are but cogs in the greater communal machine, the microbial mind. This mind was bent on eating its prey, and our minds on preventing that from happening. With all of our technology, we barely defeated the microbial mind. It will be at our door again soon enough, and it will be smarter next time.

As this book unfolds, intelligent behavior will often be portrayed as a weapon intended purely for combat in the struggle for survival. This isn't completely true, needless to say, but we must ask ourselves: why did intelligence evolve in the first place? To write novels—or to allow us to exist in a world bent on our destruction? The answer to this should be clear to any student of the living world.

In his Darwinian movie *2001: A Space Odyssey,* director Stanley Kubrick explored the evolution of intelligence. The transition from ape to man is marked by the ape's discovery of weapons to kill his fellow ape. Later in the movie, the ultimate struggle for survival takes place between astronaut Dave Bowman and HAL 9000, his spaceship's intelligent computer. Bowman loses every chess game to HAL, but when a supernatural force pits the two intellects against each other in a life-or-death battle, Bowman wins and kills the machine. To Kubrick, real intelligence has nothing to do with winning esoteric chess games and everything to do with devising better ways of killing your enemy.

Beryl Markham, in her remarkable autobiography *West with the Night*, describes how she first used her airplane to hunt elephant. The first female pilot in Africa, Markham would fly over the African plains and radio back the location of bull elephants with large tusks. Ivory hunters, on foot, would then go in for the kill. Unfortunately for Markham's young business, the elephants quickly learned the sound of her engine. They crowded together and lowered their heads to hide their tusks as soon as she arrived. Not only did they learn that hunters always followed the plane, they also deduced that the plane was only interested in big tusks. Such intelligence may be obvious only to those who kill things for a living. Like hunters.

Or physicians.

The golden age of antibiotics may be nearing its end. We gained a temporary supremacy over bacteria, but so far their ingenuity appears superior to ours in the molecular arena. Bacterial DNA has united against us and formed a global entity, an infinitely malleable device that can communicate, and think, across continents. If we step away from the world of bases and mRNA and ribosomes and look at the big picture, we can't deny the brilliance bacteria have displayed over the last half century.

Antibiotics aren't mystical things. They're three-dimensional molecules built to mimic other three-dimensional molecules, tiny structures designed to confuse bacteria. They are monkey wrenches tossed into the bacteria's internal clockworks. As quickly as we jam them in, bacteria calmly extract them. We can confuse the beast, but only briefly.

Bacteria are defining the limits of our intelligence in the biochemical sphere. We can speak, write, do arithmetic, write computer code, and even light giant metal candles and plop ourselves onto other celestial bodies, yet none of this impresses a hungry E. coli. We can't talk, think, or program an E. coli to death. To deal with E. coli, we have to exit our world of differential equations and Shakespearean sonnets and enter their world of genes and protein stereochemistry (the study of a protein's three-dimensional structure). We can only harangue them in their own language with our pathetic macromolecular diatribes. But chemistry is *their* language, and they've been speaking it for millions of years. We've known the tongue for only fifty years. We're foolish to think we can defeat microbes permanently on their home turf.

Observe now the bias of the Turing test. It's rigged to make brains look smart and bacteria look dumb. Redefine the language, and the reverse is true—we look dumb and they look smart. Well, I should say our neural brains look dumb. Our immune system, the inner brain charged with halting marauders like E. coli, happens to speak their language as well as they do.

In the end, Tom wasn't saved by antibiotics. As any infectious disease specialist knows, antibiotics never cured anyone. Our immune system ultimately cures us; antibiotics only help. The next chapter addresses the intelligence of the immune system, an intelligence that rivals that of the brain. Studying the immune system gives us an even deeper look into the fundamental nature of biological intelligence: the intelligence of the community, the intelligence of living networks.

2   THE IMMUNE INTELLECT

No man was ever yet so void of sense
As to debate the right of self-defense.

—Daniel Defoe, "The True-born Englishman," 1701

Tom and his defeated bacteria retreated from their conflict and began preparing for future encounters. The last chapter illustrated how bacteria deal with new antibiotic threats. Tom's new strategy was to quit smoking and consider (finally) getting a flu shot. Although flu shots have no direct effect on bacteria, they prevent the viral infections that serve as breeding grounds for bacterial pneumonia.

For years, despite his doctor's strident advice, Tom was convinced that flu vaccines only made people sicker, and he stubbornly refused to submit to them. But his month spent tethered to a mechanical ventilator greatly altered his attitude toward preventive medicine. Tom relented and got the shot. It proved far easier than giving up cigarettes.

Vaccines teach our internal defenses a lesson in the anatomy of microscopic pathogens. Vaccinations expose us to dead or weakened versions of dangerous microbes; they are mock infections, analogous to war games in which soldiers fire blanks instead of bullets. They prepare our military for battle without exposing our body's soldiers to real risk.

Every pathogen—viral, bacterial, or otherwise—has a unique shape, just as every person has a unique body and facial contour. Our visually oriented brains learn to distinguish people by physical appearance; likewise, our internal defenses learn to recognize microbes by their three-dimensional shapes. During World War II, submarine commanders and fighter pilots carried handbooks illustrated with the silhouettes of enemy warships, so that they could better identify friend from foe. Vaccines provide our defenses with the silhouettes of enemy microbes for the same reason, namely, advance intelligence.

It takes our defenses about fourteen days to commit the shape of a pathogen to memory and customize a defense against it. Consequently, a first-time invader may roam our bodies unchallenged for two weeks or more; that's why severe viral infections last at least that long. If unrecognized, microbes can pass under our radar for weeks and have a free hand to wreak havoc. If, on the other hand, we already know a pathogen's shape because of prior vaccination or previous encounters with the same organism, the invader encounters stiff resistance immediately and can be rejected before any infection takes hold.

So far, this is routine stuff. Most people know what vaccines are and how they're supposed to work. But let's probe a deeper mystery: what part of the body learns the lessons taught by vaccines? How do animals remember the intricate silhouettes of microbial invaders, recalling them many years later? The ability to store and retrieve large sets of three-dimensional data requires some form of sophisticated long-term memory—a type of memory typically associated with brains or advanced computers. Vaccines train neither brains nor computers. Nevertheless, the information they impart must go somewhere. But where?

The more we learn about our mysterious memory for pathogens, the more remarkable it appears. For instance, pathogen memory is very long-term. A tetanus vaccine administered in infancy protects us well into our ninth decade. Many neurological memories won't last as long. Moreover, the body's capacity for storing microbial shapes is both mammoth and precise. We can store and recall a million unique molecular patterns. Our bodies can also tell the difference between two protein molecules that differ by a single amino acid. If our eyes were equally discriminating, we would be able to distinguish two golf balls differing by a single surface dimple.

Though our memory for microbes equals or surpasses neurological memory in many aspects, the cranial brain has no role in this form of data acquisition and storage. A headless animal can still be vaccinated. Microbe memory isn't in our heads, but instead resides in a different kind of brain entirely, an alternative form of living computer called the *immune system*. The immune system is a vast, formless organ composed of billions of white blood cells dispersed throughout the tissues of the body. This "immune brain" may be the equal of the one within our skulls, yet it speaks no Turing language of words, numbers, or symbols. Instead, it converses in the language of bacteria and other microbes, a syntax composed entirely of molecular words. The immune system is our answer to the global microbial mind; it serves both as translator and bodyguard as we travel in the microscopic realm.

Like the microbial mind, the immune system must do more than store and remember discrete facts; it must also extrapolate from past experience and react to novel situations using that experience as its guide. For example, Edward Jenner hit upon the idea of vaccination by observing that milkmaids who had been infected with a mild virus known as cowpox were rendered immune to a much more dangerous illness: smallpox. By inoculating people intentionally with cowpox, Jenner could provoke an immunity to smallpox and one of the terrors of the earth, the smallpox virus, would soon verge on extinction. (The word *vaccination* itself comes from the Latin *vacca*, a cow.)

The immune system learns patterns, then uses pattern recognition to craft intelligent responses to novel threats. The cowpox virus looks *almost* like the smallpox virus, but not exactly like it. A stupid device would consider them separate organisms, but the immune system, like any intelligent pattern-recognizing machine, sees enough of a similarity to make a connection between the two viruses. Like the boy who fears all dogs because of his experience with one breed, the immune system fears all pox viruses after being exposed to one species. The immune system, like the boy, must extrapolate from past experience to survive in a world filled with predators of all shapes and sizes.

In the billion-year arms race between the immune system and the microbial world, bacteria acquired a power of reason that they wield against us every minute of every day. The immune system had to match this cognitive power, and it did so in a rather ingenious way. But before we can fathom how (and why) the immune system does what it does, we must travel back in time, back to the era of the primordial soup and the first appearance of multicellular creatures on this earth. Along the way, we must touch briefly upon the nature of life itself.

What unifying force drives the behavior of all living systems? Pose this question to a dozen biologists and the same answer will emerge: survival. In the mainstream view, life exists solely for its own mindless perpetuation.

This view strikes me as needlessly nihilistic—and, on deeper analysis, wrong. One look at a peacock or an orchid tells me that life must seek some purpose beyond relentless fission. I propose that we view life's motivation in a different way and assume that living things seek not to survive, but to *socialize*. In this context, I'll define socialization as the desire of individuals to merge into a cohesive group displaying emergent behaviors. (Recall that emergent behaviors arise almost magically from the synergistic interaction of a society's members. For example, the V. fischeri form a society that knows when it's inside a fish. That knowledge, which goes beyond the powers of any single bacterium, becomes an emergent property of V. fischeri colonies.)

A particularly stunning example of emergent behavior is human consciousness. I know I promised not to bother with consciousness, but I'll bend that rule temporarily. Consciousness "emerges" from billions of neurons connected to one another in exactly the right way. Although we now grow human neurons in a test tube, we can't create a brain simply by lumping a few pounds of mass-produced nervous tissue into a gelatin mold. The brain may look formless, but it's really an intricately woven society. To display true emergent properties, a society—whether built from bacteria, neurons, or human beings—must become more than the simple sum of its parts. To yield an aware person, the social structure within the human cerebrum must follow a defined architecture, and it is this architecture—the "wiring diagram" linking the constituent neurons together—that provides the blueprint for the neuronal society and defines the emergent properties arising from a brain. In a general sense, life, intelligence, and consciousness are all emergent social properties springing from the architecture of a large group. In V. fischeri, the architecture is a fairly simple one:

bacteria linked together by a single hormone. In the brain, the density and scope of the interconnections boggle the imagination. The differences among living systems turn out to be differences of degree, not of principle.

For an emergent society to arise, members must communicate with one another, but that doesn't mean that all members have to cooperate. In fact, even if many members exist in a state of mutual war, the society can still function cohesively. Cybernetic models of social behavior define competition as negative information flow between members, and cooperation as positive flow. In the brain, neurons competitively inhibit other neurons during the learning process, as will be seen in chapter 7. The distinction between competition and cooperation becomes purely mathematical, devoid of all metaphysical overtones. The dynamic interplay of positive and negative forces stabilizes a real society and helps craft emergent behaviors. The immune system is a case in point, as will be seen shortly. As one of my physiology professors liked to point out, the statue of David was created by inflicting purely destructive forces onto a piece of square marble. Sometimes beauty and order arise out of conflict and death.

In the social paradigm of biology, the story of life on earth has a common plot line: smaller things aggregate into larger communities manifesting emergent properties. The property of being alive is itself emergent. During life's genesis, small molecules coalesced into organic macromolecules, which then formed metabolic pathways. These pathways linked to create living cells, and living cells assembled into organs. Organs made organisms, organisms made ecosystems, and, at last, ecosystems united into a living planet. From the first condensation of macromolecules in ancient clays to the modern amalgam of human minds on the World Wide Web, life forever connects and expands.

We could say that the need to socialize is in our genes, but

that wouldn't be true. The need to socialize *predated* the existence
of genes; genes themselves are a product of the socialization
process. Genes simply keep track of how things like to organize,
serving as a form of social register for enzymes (more on this
later). The drive to socialize gives life a motivation above and beyond mere survival. If survival was life's only goal, why didn't we
cease evolving at the bacterial stage? Bacteria excel at surviving;
in truth, they appear far more adept at it than we are. They can
live almost anywhere, from hot boiling springs to the ocean's
frigid depths, and can eat anything, including hydrogen sulfide
and sunlight. Humans are fastidious wimps by comparison, hardly
worth the eons it took to spawn us. Developing complex structures like mammals and flowers seems a stupendous waste of time
and effort if staying alive were all that mattered. But life doesn't
just survive; it grows larger and more intricate, always manifesting
new properties.

Socialization has been life's chief guiding principle from the
beginning of time. Organic molecules participated in nonliving
social groups for millions of years before they finally generated
life as we know it. In those prebiotic days, survival didn't matter,
because if you aren't alive, you don't need to survive. There is no
"chicken or the egg" dilemma here: the need to socialize came
first, the drive to survive came much later.

But enough of this. Let's return to the original topic: self-
defense.

Not long after macromolecules assembled to form the first
free-living cells, cells began coalescing into the first multicellular
organisms—the need to socialize did not end with the generation

of life itself. In turn, cellular socialization leads inevitably to the need for defensive capabilities.

Forming cellular societies has its advantages; there has always been safety in numbers in a hostile world and, at first, primitive cellular societies welcomed all newcomers. However, these naive country clubs soon realized that indiscriminate membership policies are counterproductive. Many bacteria would love to join a V. fischeri colony and reap the food rewards of a fish hunt, but if new members don't glow like V. fischeri, they'll quickly dilute out the bioluminescence that gives the colony its competitive edge in the first place. To guard its advantage, any colony must learn to identify and eject parasites.

Cellular colonies, like all societies, had to set rules: those that eat the colony's food or reap its protection must also do the colony's work. But rules are no good without enforcement. Consequently, violence emerged early in life's history: good members were accepted and interlopers destroyed. The lethal distinction between "us" and "them" came into being. To enforce this distinction, colonies became gated communities and developed techniques for policing their borders and scanning their memberships for freeloaders, which were then excluded or killed outright.

The multicellular colony's first line of defense is a structural barrier. Examples of structural barriers include vertebrate skin, the chitinous shells of beetles, the bark of trees, and the gelatinous exudates surrounding spineless invertebrates. These barriers are castle walls and moats that physically block an invader's entry. For many organisms, including most higher plants, barriers provide nearly all the defense they'll ever need. While physical barriers can be breeched by injury, immobile organisms like trees rarely suffer major trauma.

But external shielding alone won't do the job for all creatures.

In some cases, shielding is impractical; it would be difficult, for example, to enclose an amorphous V. fischeri colony in a protective container (although the host fish may serve this purpose to a certain degree). In mobile creatures, the need for flexibility places an upper limit on the strength of their external housing (imagine running from a fleet predator while encased in hard tree bark). Even heavily armored brutes, like the Goliath beetle, must have vulnerable joints in order to move their limbs freely. Moreover, armor can be quite heavy, limiting an animal's endurance and speed. In the end, no matter how strong the barrier, a predator's teeth will eventually prove stronger.

Constant movement over treacherous terrain puts an animal's barriers at risk for injury, and because they can't photosynthesize their fuel like their plant cousins, animals must periodically lower their shields to allow food to enter, which then makes them vulnerable to invasion though their relatively flimsy gastrointestinal systems. Many human pathogens cause disease only after they are ingested: polio enters the body through the intestine, as does the bacterial species causing typhoid and cholera. For these and other reasons, structural barriers, although simple and effective, won't suffice for all organisms. In a structural defense, the distinction between "us" and "them" (in immunology, the distinction between "self" and "nonself") becomes strictly geographic. "Us" is whatever lies inside the barrier, "them" whatever lies outside. As a result, this form of defense requires no intelligence at all. To go beyond structural defenses, organisms had to become considerably smarter.

The next version of defense to evolve was *non-adaptive immunity*. In non-adaptive immunity, organisms secrete substances toxic to invaders but to which they themselves are immune. This type of defense is considered non-adaptive because it can't be customized by individual creatures to target a specific invader. Organisms armed with non-adaptive immunity own a powerful, but inflexible, arsenal. When fending off penicillin-sensitive bacteria, a penicillium mold makes penicillin. When faced with a penicillin-resistant bacterial strain, the mold also makes penicillin; it's unable to make anything else. The mold can't adapt to a new situation, at least not in its own short lifetime. Over many generations, molds can genetically alter their antibiotics in response to evolving bacterial resistance, but such adaptation occurs slowly and only in the species as a whole, not within individual colonies.

Non-adaptive immunity is more intelligent than structural immunity, but barely so. A physical barrier is simply replaced with a chemical one. The main problem with non-adaptive immunity is that it's, well, so non-adaptive—or at least non-adaptive within individual creatures. The average house fly produces about fifteen different antibacterial toxins; although these agents can kill a broad spectrum of bacterial species, they can't kill them all, and the genetic genius of bacteria allows them to circumvent any toxin eventually. If an isolated fly encounters a bacterial species with fresh resistance to its limited repertoire of bacterial toxins, our poor insect friend will find itself defenseless and out of luck. A single fly, like the single mold colony, lacks any capacity to alter its toxins in response to evolving bacterial resistance. We can't vaccinate the fly, either—it can't learn anything new. An insect emerges from the egg with the same defenses it will have at death.

Fortunately for the fly as a species, the inflexibility of its individual defenses matters little, because insects share with bacteria and molds a short generational time and the power of numbers. While the single fly can't modify its toxins in response to bacterial resistance, millions of evolving flies can. Consequently, the term "non-adaptive" is somewhat misleading here; non-adaptive immunity *does* adapt, but it adapts at the species level over many generations, and many years. *All living systems learn, but not necessarily at the same pace or in exactly the same way.*

The immune systems of isolated flies can afford to be stupid. Like bacteria, flies distribute their defensive intelligence over unfathomable numbers of organisms. Simple problems are addressed by individuals, while more complicated dilemmas can be left to Darwinian selection at the level of the species. The fly, as a species, must possess an immunity every bit as smart as our own; otherwise there would be no flies. Although flies deploy their defensive brainpower differently than mammals, the intelligence must still be there, dispersed throughout the fly species as a whole. I'm reminded of the joke about two hunters who are suddenly confronted by a vicious bear. One hunter turns to run, but the other hunter warns: "Wait, you can't outrun a bear!" To which the first hunter replies: "I don't need to outrun the bear, I only need to outrun you!" All multicellular creatures face a common predator: microbes. If one species proves much slower than others in running from this threat, it will soon be annihilated.

The last chapter illustrated the brilliance of modern microbes. The microbial world is a dynamic evolutionary engine, a vast sea of interconnected nucleic acids capable of adapting to any challenge quickly. A species that lags behind microbes in intelligence will soon be devoured to extinction. In this context, the advent of vertebrates (creatures with rigid spines) meant defensive schemes had to change further. Compared to the typical invertebrate, vertebrates reproduce in tiny numbers and live for many years, even many decades. Vertebrates don't lay a thousand eggs at a time, nor can they wind through a dozen generations in a single year. From a microbial perspective, vertebrate organisms evolve at a glacial rate.

Like all living things, vertebrates are at constant war with bacteria. To keep from rotting alive, they must ceaselessly kill or neutralize a vast army of predatory bacteria, viruses, fungi, and other pests. Given the differential rate of adaptation, microbes versus vertebrates, the microbes should defeat vertebrates handily, a case of the hare versus the tortoise in a genetic race for survival. Flies can go toe-to-toe with bacteria by churning out new genetic variations with near microbial rapidity, but vertebrates can't manifest an equivalent genetic plasticity.

Or can they?

To compete on an equal footing with microbes, newly evolving vertebrates could not rely on large-scale evolution, acting on many generations, to solve their thorniest problems. Individual animals would have to shoulder more of the intellectual load and do a little "evolving" on their own, and do it in the course of a single lifetime. Does this mean that vertebrates abandoned the principles of Darwinian evolution? No, they simply internalized the process, making it portable within individual animals. And like their enemies, the microbes, they also supercharged it a bit.

This idea sounds bizarre. Supercharged evolution housed inside a single animal? After all, evolution implies mutation or some other form of permanent DNA modification. For individual vertebrates to "evolve" genetically during their lifetimes, they would have to alter their own DNA intentionally as they age. Are the genes I possess now different from those I inherited at birth?

Yes. Well, at least some of them are.

Evolution also means bloodthirsty competition, a mortal battle to survive that leaves only the fittest genetic variants standing. Are we really at war within ourselves? Is there a "survival of the fittest" showdown going on inside me as I write this? Yes again.

Bizarre as the idea of internal evolution sounds, it's quite true. Genetically, I'm not the same person I was forty-five years ago. I *have* evolved—or, more accurately, my immune system has evolved—to keep pace with the genetic plasticity of the microbial world laying siege to my flesh. In addition, the white cells in my body are fighting for the right to survive, as certainly as the squirrels of the forest or the fish in the ocean compete to prove their own genetic superiority.

Hours after he received his flu vaccination, Tom began creating entirely new genes, his white cells mutating in an attempt to find a novel way of neutralizing the virus, just as bacteria mutate in order to discover some novel way of neutralizing an antibiotic. Mutant white cells then began competing with each other to determine which ones recognized the viral proteins with the highest accuracy. The fittest mutants thrived; the losers dwindled. Tom became resistant to the flu by randomly churning out anti-flu genes and letting those genes duke it out. The best genes would become part of his permanent DNA library; the lesser genes would be discarded.

Vertebrate immune systems, like the microbial mind, have learned to harness and accelerate the power of genetic instability and natural selection. They exploit their own white cells by putting them under stress and making them adapt—or die. Our bodies artificially create a competitive cauldron and reap the benefits of the victors. Remember the science fiction story in the first chapter? As it so happens, we exploit our white cells as ruthlessly as the scientist exploited his tiny beings. In fact, this is a common theme among intelligent biological systems: create an artificial ecosystem, make individuals compete for some scarce resource, and then reap the benefits of the Darwinian adaptation that ensues. (As will be seen in chapter 7, the cerebral cortex is a miniature ecosystem wherein neurons compete for electrical information.)

Vertebrates use a complicated form of defense known as *adaptive immunity*. In contrast to the limited arsenals of a non-adaptive immune system, the adaptive immune system crafts a unique response to each new infection. A fly has fifteen anti-microbial toxins; we have a million, each targeted to a specific microbial molecule. A vertebrate's "toxins" are called immunoglobulins and T-cell antigen receptors (TCARs), giant proteins that are tailored to recognize and attach themselves to invading microbes. To keep things as simple as possible, I'll refer to all such tailor-made toxins by an older and more generic term: *antibodies*.

Antibodies serve as death markers, binding tightly to invaders and identifying them as objects to be destroyed. Lumberjacks paint red Xs on trees to be felled, fighter pilots cast their laser light on a target to be bombed, and antibodies attach themselves to invaders. In the immune system, the destruction doesn't fall to chainsaws or missiles, but to killers known as phagocytes (literally "eating cells"), single-celled scavengers that roam the body in search of anything marked as food by antibodies.

A foreign molecule—one that prompts antibody formation— is called an *antigen* (short for antibody-generating molecule). Antibodies bind to an antigen by precisely molding to the antigen's physical contour, much like a key fitting a unique lock. Also like a key, the antibody protein consists of two parts: (1) a constant region (which is shared by all antibodies), and (2) a variable region uniquely contoured to fit a particular antigen. On the key, the constant region would be the place we grab and turn, the part that looks similar on any key. The key's serrated edge forms its variable region, the part that gives a key its unique identity.

An antibody's variable region (also called its *V-region*) binds to only one type of antigen, just as a key's serrated edge opens only one lock. The V-region is also known as the antibody's "binding site," as it's the region that binds tightly to the antigen. The rest of the antibody molecule is constant and shared by all antibodies. This constant region also serves a common purpose: summoning killer phagocytes. To destroy an invader, antibodies attach themselves to the invader's surface like lampreys on a shark, their constant regions dangling behind them like the lamprey's tail. The constant regions call the phagocytes to supper.

The key analogy, although easy to understand, isn't really all that structurally accurate. Antibodies are globular, like bowling balls, not linear like keys, and the antigens they bind are also three-dimensional. In a real antibody, the V-region binding site is a negative 3-D image of the antigen to be recognized. Consequently, the following analogy might be better:

Take ten round balls of clay, each six inches in diameter. Now take ten small toy trinkets (for example, the tokens in a Monopoly game—the shoe, the dog, the iron, and so on). Press each trinket into one ball of clay, then remove the trinkets and bake the balls in a drying oven. We now have ten hard balls of clay that look almost the same, except that each now has a small pocket bearing the negative 3-D image of a particular trinket—the trinkets are the antigens, the balls are the antibodies. The small pockets form the variable regions, while the remainder of each ball's mass forms the constant regions shared by all ten balls. Because the balls are large and the trinkets small, the constant regions dwarf the variable regions. Likewise, in real antibodies, the V-regions are only a small part of the larger antibody molecule.

The V-regions are the heart of the immune system's specificity. If any antibody V-region recognizes a target, it will be attacked. Conversely, if all V-regions fail to recognize a target, it will be ignored. At birth, the immune system is pretty much a blank slate, equipped with randomly generated V-regions that don't recognize anything in particular. During our lifetimes, new V-regions are crafted to recognize each new pathogen as the need arises.

We are now faced with a million-dollar question: how does the immune system create a custom-fitted V-region for every new antigen it encounters? Several weeks after Tom got his flu shot, his bloodstream contained antibodies with V-regions targeted to influenza antigens. Tom could now manufacture an entirely new class of antibodies designed to recognize viral proteins. How did his body accomplish this remarkable feat of protein engineering, and in just two weeks?

It's tempting to speculate that we have a reservoir of claylike antibodies that can be molded to fit new antigens, but that's not how proteins work. The shape of any protein, including antibodies, is determined by its unique amino acid sequence; that sequence, in turn, is determined by a DNA base sequence contained in the cell's chromosomes. Thus, a protein's shape is imprinted in our genes and can't be significantly altered once it has been made; proteins aren't clay. To change a protein's shape, the genetic code that dictates its assembly must be modified first. To make a new antibody, you have to create a new antibody gene. For decades, immunologists struggled to understand this enigma. Yes, we continuously make new antibodies, but does this really mean we continuously make new antibody genes as well? And if we do, how does the body generate these genes? More important, how does our DNA "know" where to modify itself in order to generate a new protein that just happens to be an exact negative image of some foreign molecule? This sounds impossible. The immune system would have to be like a musician who can directly imprint music onto a compact disc without hearing or playing it first! And one who could write it *backwards* (remember, the V-region is the *negative* image of an antigen).

In the context of a Turing test, the immune system looks too intelligent, almost like a magician. We inject an influenza protein into the body and, several weeks later, out pops an antibody that recognizes that protein with uncanny precision. It accomplishes this feat at the *genetic level,* so that our ability to recognize influenza antigens becomes part of our DNA. As we grow older, our antibody repertoire expands exponentially as every new infection, new food, new grain of pollen, and new vaccine spurs our immune system to make new antibodies. By the time we die, the immune system has manufactured well over a million different antibody molecules (or, to be more precise, over a million different V-regions). That's a pretty big set of keys.

And each new V-region must have its own new *V-gene* somewhere in our DNA. In trying to piece this puzzle together, immunologists were faced with a rather impossible choice: either (1) we are born with all the V-genes required to create every possible antibody we will ever need, or (2) we continuously create new V-genes during our lifetimes via mutation or some other genetic trick. The first choice turns out to be physically impossible. If a human egg contained a V-gene for one million or more possible V-regions, it would be the size of a tennis ball.

The master copy of DNA needed to make one organism—the DNA contained in the fertilized egg—is called our germline DNA. No vertebrate germline DNA can carry that many V-genes; it just isn't large enough. Besides, this scheme makes no sense defensively: if our antibody repertoire were simply frozen in our germline genes, we'd be no better off than a house fly. While we might have a larger number of toxins than the fly, thanks to our larger chromosomes, the immune system of any individual would still be inflexible, predetermined at birth.

But the second choice—the notion that our DNA somehow changes inside our bodies to create new proteins—also seemed unpalatable. Biologists liked to believe that a pristine germline carries all the genetic information we'll ever need from cradle to grave. For individual animals to change genetically during their lifetime sounds too dangerous. After all, cancers arise because of alterations of our germline DNA, so surely we wouldn't toy with our DNA blueprints intentionally.

Ah, but indeed we do. In the 1970s, improved DNA analytical techniques allowed immunologists to peer into the DNA of vertebrate immune systems, and they soon learned that we *do* evolve genetically in response to external threats in the same way that bacteria and flies evolve. We now know that our immune system employs the identical methods of gene-shuffling and directed hypermutation that allow the bacterial world to defeat our best antibiotics. The immune brain, like the microbial mind, is a Darwinian computer of immense power. We must now delve a little more deeply into the mechanism of antibody creation to see how natural selection operates beneath our own skin.

Antibodies are manufactured by a specific class of white blood cells known as lymphocytes. Each mature lymphocyte manufactures only one type of antibody at a time, although the nature of that antibody may change during the lymphocyte's life span. Unlike bacterial cells, vertebrate cells like lymphocytes can splice two or more genes together to make a single protein molecule. To build an antibody protein, the lymphocyte uses a "C-gene" to produce the constant region of the antibody and a separate set of

V-genes to make the all-important variable region. The two parts are then incorporated into a final protein. To complicate matters further, more than one V-gene contributes to making a single V-region.

During our fetal development, immature cells destined to become lymphocytes must choose the general nature of their antibody's V-region, similar to college freshmen deciding on their majors. In humans, fetal lymphocytes can select from more than a hundred V-genes stored in the germline DNA. To generate maximum antibody diversity, the immune system uses an ingenious trick: the V-region isn't made from just one V-gene, but from several genes grafted together. The fetal lymphocyte can randomly mix and match V-genes to come up with any sort of antibody it desires.

Consider this analogy: assume that a germline V-gene is a playing card and that each antibody V-region is a poker hand consisting of five V-gene cards. The germline is the "deck." During its fetal maturation phase, a lymphocyte deals itself a V-region at random from the germline. The number of unique V-regions that can be generated by this random-dealing method becomes huge, on the order of 100 trillion or more. By letting lymphocytes mix and match a limited number of germline V-genes to make different antibodies, the immune system generates an enormous range of V-regions using only a tiny amount of germline DNA. It's really a beautiful system.

Of course, not all V-gene "hands" are winners. By definition, the immune system must attack only foreign antigens, leaving our own cells and molecules alone; but, since the process that generates our initial antibody diversity is random, it's inevitable that antibodies directed against our own molecules will arise. These harmful *auto-antibodies* are deleted from the antibody repertoire early in life, rendering the immune system tolerant of the body in which it resides.

We can now see how a tiny amount of germline DNA can yield over a million V-region "keys" with which to turn the antigen "locks" of any potential pathogen. But that still doesn't explain how a given antibody ends up fitting a foreign antigen so exactly. So far, the immune system still doesn't look very smart. It can grind a lot of keys, but it does so by chance alone, not caring what locks they fit—so long as they're not friendly locks.

By exterminating the unlucky lymphocytes that decided unknowingly to make auto-antibodies, we enter the world armed with antibodies directed only against foreign things. But that's about all we can say about an infant's immune system: it recognizes things other than the infant's own tissues. Since our very first antibodies are random creations dealt by chance from germline V-genes, an antibody that recognizes a specific influenza virus would do so by purest accident.

Armed only with random antibodies, the infant's immune system is indeed a tabula rasa. It may contain over a million kinds of antibodies, but none has been tailored to recognize an actual microbe. An infant's immune system is like a recent college graduate: full of vigorous potential and theoretical knowledge, but sadly lacking in any real-world experience. Were it not for the more worldly maternal antibodies transmitted to the fetus via the placenta, a fragile infant would soon succumb to infection. Even with mom's help, the average child experiences a baker's dozen viral infections in the first year of life. My wife and I, like all parents, have lost a great deal of sleep waiting for our children's immune systems to shed their infantile stupidity and start doing their jobs.

We now know how the immune system makes a million antibodies with only a handful of germline genes, but we're left with

yet another mystery: how does the infant's random antibody repertoire grow into the educated repertoire of an adult? My forty-five-
year-old immune system recognizes smallpox, influenza, polio—
and, unfortunately, ragweed pollen—with startling precision. I
wasn't born with this immunity, so how did I acquire it? How does
the immune system learn new antigens and imprint the memory
of their shapes into our very genes? This is where Darwin enters
the picture again.

When viewed through a microscope, all lymphocytes look the
same: little round cells with big nuclei. But because different lymphocytes have, at random, chosen to mix and match their antibody
V-genes differently during their maturation phase in utero, lymphocytes differ genetically from one another. Consequently, an
adult lymphocyte and its progeny form a unique *species* of white
blood cell. The immune system is really an ecosystem containing
thousands of genetically distinguishable lymphocyte species competing for the same finite resources, and, as in any zero-sum game,
some species thrive at the expense of others.

This sounds very bizarre. We're used to thinking of organ systems as vast collections of cells with a common purpose, all cells
marching lockstep together and sharing resources equally. We
don't picture our liver cells as competing with one another to survive—and they don't. The liver is a kind of Marxist commune:
from each cell according to its abilities, to each according to its
needs. But the immune system is not the liver, for inside the immune system it's capitalist war. Each lymphocyte species seeks to
expand infinitely, even if it means killing all other lymphocytes.

Let's stop for a moment and consider this carefully. We tend to
think of our cells as having altruistic motives above and beyond
their own selfish needs. A bacterium floating in a pond thinks only
of itself. But shouldn't my lymphocytes have to think of my needs

first? Ironically, they do *not*. They neither know nor care that they're only cogs in the greater machine of the body—they wish only to perpetuate their own genetic lineage at all costs. As noted earlier, a large group can still function cohesively even when all members compete to achieve their selfish ends. This will be a recurring theme: intelligent behavior can emerge out of groups of very selfish beings.

To grow and thrive, lymphocytes must be exposed to an antigen that fits their unique V-regions. In the presence of the right antigen, lymphocytes can go hog wild, making new lymphocytes and pumping out antibodies all day long. In the absence of antigens, a lymphocyte can only sit quietly in some lymph node and wait, like Tennessee Williams' Laura in *The Glass Menagerie*, longing for the right gentleman caller to arrive.

A lymphocyte's surface is studded with antibody molecules that are wired into the lymphocyte's internal growth mechanisms and act like hormone receptors. The appropriate antigen acts like a growth hormone, triggering the lymphocyte to divide. Without antigen stimulation, a lymphocyte lingers, childless and forgotten, and may even die out completely. In the lymphocyte's ecosystem, antigens are prey and each lymphocyte species may "eat" only a certain kind of antigen food, due to the limited specificity of its V-region binding site.

In a macroscopic ecosystem, the landscape is populated with different species with specific dietary requirements. Rabbits eat grass, wolves eat rabbits, bacteria eat dead wolves. The relative number of each species is determined by the abundance of their respective food sources. Likewise, the distribution of lymphocyte species is driven by the availability of their respective antigens. By coupling a lymphocyte's growth and survival to the availability of its target antigen, the immune system sets the stage for a

Darwinian competition designed to identify which lymphocytes are best at recognizing real invaders. How ingenious!

Let's return to Tom. Even in his eighties, Tom's immune system retains many randomly determined V-regions dealt out during his time in the womb. After he was inoculated with flu vaccine, viral proteins entered Tom's bloodstream, where thousands of lymphocyte species began sniffing out these proteins, trying to recognize them. Like Cinderella's evil stepsisters jamming their feet into the glass slipper, an army of lymphocytes tried forcing their V-regions into the nooks and crannies of the viral proteins, hoping that a precise fit would occur and that they might start proliferating. At this stage, the fit between antigen and antibody is pure trial and error; perfect matches will be unlikely, but cruder matches will be common.

For the evil stepsisters, the outcome was an "on-off" sort of thing. Either they fit the shoe and got the prince or they didn't. For lymphocytes, the relationship between antigen-fit and cell stimulation is more gradual. If the V-region binds to a new antigen a little bit, the cell gets stimulated to divide a little bit as well. On the other hand, if the fit is precise, the stimulus to divide becomes great. Those lymphocytes that, by accident, recognize the antigen exactly will eventually edge out less fit competitors. The degree to which an antibody's V-region fits an antigen is called its *affinity*. An exact fit means a high affinity; a poor fit means a low affinity; no fit at all means zero affinity. During an infection or after vaccination, lymphocyte species proliferate in proportion to their affinity for the new antigens.

In the feeding frenzy that follows the infusion of fresh antigen, lymphocytes that possess antibodies with low affinities for the antigen will be mildly stimulated. As they begin to proliferate, they start mutating their V-genes in an attempt to yield a daughter species with an even higher affinity for that antigen. At this point, the process escalates into a classically Darwinian competition: mutants square off against other mutants in response to an influx of antigen food. Those with the highest affinity will soon bind up most of the food, leaving little for their weaker competitors. Eventually, only the lymphocytes with the highest affinities—the best achievable match between V-region and antigen—still proliferate. This competition is so effective that mutant antibodies with enormous affinities for any new antigen will quickly emerge from the most random pool of V-regions.

Occasionally too little antigen is introduced, or the antigen is eliminated too quickly for optimal fine-tuning of antibody affinity. The best antibodies arise from repeated exposures to antigen. Practice makes perfect, which forms the impetus for "booster" shots. In childhood, we receive vaccinations against diphtheria, tetanus, and pertussis (DPT) three times, each spaced a month apart.

In the introduction to this book, I described the concept of pattern recognition. In the immune system, pattern recognition is embodied in the relationship between antibody affinity and lymphocyte stimulation. The match between an antibody's V-region and an antigen doesn't have to be exact; an approximate fit is all that's needed to kick-start the immune system and send it in search of something better. The most effective antibodies have a high affinity for their targets, but antibodies with low affinity can still clear an antigen from our bodies; they just take longer to do so. Like all intelligent machines, the immune system doesn't need perfect input (or perfect output) to achieve a reasonable, if not ideal, outcome. Intelligent machines start out with a best guess and work from there. In a pinch, a guess may even be good enough to get the job done. In general, the hallmark of biological intelligence (as opposed to machine intelligence) is that it prefers to get *good* answers very quickly, rather than find *perfect* answers very slowly.

After all antigen has been recognized and consumed by phagocytes, high-affinity lymphocytes generated by this process will decrease in number but never return to their previously low levels. Moreover, these lymphocytes assume the special status of permanent "memory cells." The immune system is like the pro golf tour in this respect. If a new player on the men's tour goes years without doing well in a tournament, he gets booted back to the bush leagues. If, on the other hand, he wins the U.S. Open, he gets to stay on tour for most of his career, even if he never wins again. In the parlance of golf, a major tournament win grants him an exemption that permits him to stay on the tour even if his game never again ascends to great heights. Likewise, high-affinity antibodies generated during antigen exposure get lifetime "exemptions"; they will thrive even if their antigen never returns.

Unfortunately, the mutant genes created in response to a novel antigen don't find their way back into the germline. The V-gene mutations occur only in the DNA of adult lymphocytes, not in the genetic master copy contained in our ova and sperm. A mother can pass only fully formed antibodies into the bloodstream of her fetus, not the V-genes that encode them. For example, my wife is immune to polio because she was vaccinated as a child and has engineered the genes to make anti-polio antibodies. When our daughters were born, they received enough anti-polio antibodies from their mother, via the placenta, to last about three months (the typical life span of antibodies in the bloodstream). However, the children could not make these antibodies themselves, because they lacked the appropriately trained lymphocytes. To be able to make their own antibodies, they had to be vaccinated too. The maternal antibodies would protect them only until the vaccination ran its course. Unfortunately, when we die, the marvelous ecosystem of lymphocyte species within us—an immense data warehouse with its one million tailored V-genes—dies with us. Each immune system, like each brain, is a unique jewel, a record of that individual's singular experience with the outside world. We can't inherit our parents' immunity to pathogens, just as we can't inherit their knowledge of Italian operas. The mother can give an antibody to her child temporarily, but only infection or vaccination can teach that child to be immune for a lifetime.

So far, I've grossly understated the true complexity of vertebrate immunity. For starters, I've neglected the difference between humoral immunity (mediated by free-floating antibodies called immunoglobulins) and cellular immunity (mediated by lymphocytes studded with T-cell antigen receptors). Humoral immunity works well against single molecules and simpler organisms like viruses. Cellular immunity, on the other hand, targets more complex structures—the rejection of transplanted organs is the best example of cellular immunity. The lymphocytes responsible for humoral immunity are called B cells (named for a structure in chickens known as the bursa of Fabricus), while those responsible for cellular immunity are called T cells (T for the thymus gland in the neck, where the cells mature).

Most cancers of the immune system (the leukemias and lymphomas) arise among B cells. At the opposite extreme, the disease known as AIDS is caused by viral execution of T cells. In particular, HIV kills a subspecies of T cells, the T helpers, which assist B cells. There's yet another class of T cells, the T suppressors, that inhibit B cells. The genetics of T cells and B cells also differ. For example, TCARs undergo genetic rearrangement in humans, but not mutation. Complicated enough for you? It gets worse.

During the battle to recognize a new antigen, lymphocytes not only compete with each other to survive, they also form complex alliances. Populations of lymphocytes (usually B cells and T helper cells) chatter among themselves using hormonal signals known as cytokines, which include the famous interferons and interleukins, substances that made headlines a decade ago as anticancer weapons. The situation is muddied further by *idiotypic* interactions among lymphocytes. During idiotypic interaction, one species of lymphocyte attacks another directly at its antibody binding site. Thus, not only do lymphocytes compete by vying for a limited antigen supply, they also fight it out *mano a mano*. The concept of idiotypic interactions is a bit too complicated to address in any detail here. Suffice it to say that all lymphocytes in the immune system interact to a certain degree with each other, just as all neurons in the brain interact with each other in some way.

Here's where organ-wide intelligence may come into play: there must be some coordination among the vast numbers of lymphocytes that attack larger structures like bacteria. Lymphocytic coordination would enable the immune system to perceive whole cells, not just isolated antigen molecules. The V-region of an antibody is relatively small, able to bind only a tiny part of any antigen. The small bit of a larger antigen recognized by a V-region is called an *epitope*. To understand the difference between an antigen and an epitope, suppose we have an army of blindfolded volunteers probing the *Titanic*. Each person knows only what's accessible to his or her hands, since the ship is vastly larger than any individual. Likewise, most antigens are vastly larger than a single antibody molecule. The V-regions are like "hands" feeling the features on a larger structure. Epitopes are those antigenic features accessible to the immunoglobulin's small V-regions. If the *Titanic* were an antigen, a porthole or a piece of deck railing would be an epitope.

If the volunteers don't interact, their view of the ship will be quite skewed. However, if they compare notes intelligently, our army of ship-feelers might be able to integrate their combined information into a reasonable idea of what the whole ship looks like.

Does the immune system do likewise? Does it have the ability to "sense" immense antigens, like whole bacteria, by integrating the bits of information derived from its vast army of lymphocytes? Or are the lymphocytes simply lone wolves, solitary predators looking only to eat solitary epitopes? We don't know the full answers to these questions yet—questions I first raised over a decade ago—although we do know that the immune system is sufficiently organized for such a task. I find it unfathomable that an instrument as intricate and beautifully crafted as our immune system was built to play the tiny, solitary notes of epitopes. Rather, the immune system seems better suited for playing great antigen symphonies.

Let's now marvel at the immune system's design. Here we have a continuously evolving ecosystem, driven by genetic rearrangements and mutations, regulated at the organ level by complex patterns of cooperation and competition among millions of unique cellular species. The net result is a highly coordinated and tightly controlled Darwinian computer that learns and recalls vast numbers of 3-D images.

And we now see where immune memory resides: both in mutant memory cells and in complex patterns of lymphocytic interaction. Immune memory is an emergent property of a lymphocyte ecosystem, a vast network formed by all immune cells working either with or against each other. No single lymphocyte can generate immunity, just as no one gas molecule can define the concept of temperature. Only the immune intellect, wired as delicately and deliberately as a brain, can accomplish this remarkable feat.

In the Turing model, if we could converse with an immune system in the language of antigens and antibodies, we would sense an immense intelligence. We put an antigen into the black box and out pops a customized antibody (and corresponding antibody gene) in a matter of days or weeks. Our best supercomputers couldn't yield these results in any amount of time. As was the case with the microbial mind, when we peer into the black box, we find Mr. Darwin staring back at us.

Our bodies are truly like the protagonist of the first chapter's science fiction story. We have intentionally imprisoned a population of living beings—lymphocytes—and structured their environment so that they will produce something useful to us. They don't adapt to please us, they adapt to survive—to eat antigens so they may live to see another day. Like the miniature beings of the story, lymphocytes don't even know that we exist. They do what they have to do, and we harness their selfishness to suit our needs. Vertebrates have engineered the lymphocyte ecosystem so that Darwinian behavior becomes focused on a defined task of our own choosing. We farm lymphocytes and harvest their abundant crop of intelligence. That so many individuals in our artificial ecosystem must perish to achieve our desired result matters nothing at all to us. We may see the invertebrate world as cruel because it sacrifices so many flies for the common good, but we shed no tears for our lymphocytes as they fall, unwittingly, on their swords for our own survival. Large-scale selfishness exploiting small-scale selfishness—a common theme in life's saga.

Are we that much better or smarter than flies and bacteria

when it comes to self-defense? Not really, just better and smarter for our own peculiar needs. The vertebrate's immune system is a souped-up version of the microbial mind encased in hide and fur. Creatures respond to environmental challenges in similar ways: through genetic adaptation superimposed on ecological conflict. Compared to bacteria, vertebrates only package their evolutionary computer differently, housing it in one long-lived vehicle instead of parceling it out to many short-lived ones. The circuitry, however, is similar in all living systems.

The proof is in the pudding: our immune systems and the microbial mind must be equally smart, because we coexist in a steady state, a biological version of mutually assured destruction in which neither gains the upper hand. The moment our immune systems become smarter than microbes, we'll wipe them out, and vice versa. The "two worlds" are as adept as they need to be and no more.

I put the phrase *two worlds* in quotes because there only *appear* to be two distinct worlds, vertebrates versus microbes. Dispense with all of the skins and hides that artificially parse the animal kingdom and we see a unified ecosystem in which lymphocytes and microbes interact as a single society, trillions of beings—bacteria, viruses, molds, lymphocytes—locked in a state of dynamic equilibrium. They are as oblivious to our neurological world as we are to their biochemical world.

The even scarier part comes when we realize that the same argument applies to nervous systems. Immune systems and nervous systems are just different manifestations of the same invention: the portable evolution machine. What we call non-adaptive immunity is, in the neurological realm, analogous to instinct. Adaptive immunity, the vertebrate's ability to customize a quick response to any novel situation on demand, would then be analogous to neurological reason.

It can be no accident that the evolution of adaptive immunity precisely parallels the evolution of the higher nervous system. At the same phylogenetic point where an insect's ganglionic instinct gives way to the more cerebral associative analysis of vertebrates, non-adaptive immunity transforms into adaptive immunity. And for the same reason: individual deer and baboons are far less dispensable to their species at large than are individual brine shrimp and swamp mosquitoes. Nature couldn't waste vertebrates in wholesale Darwinian experimentation, so it provided animals with their own evolutionary laboratories: a nervous system to address macroscopic threats and an immune system to deal with microscopic ones.

Even among vertebrates, the diversity of the antibody repertoire is proportional to the size of the brain. Fish and birds have very limited adaptive immune capabilities, while mammals have complex immune systems equipped with virtually infinite repertoires. Why did the immune system grow in complexity in concert with the advancing complexity of the brain, given that each organ operates in different worlds (microscopic and macroscopic)? Is there, perhaps, no functional difference between our macroscopic and microscopic defensive needs?

And, if we accept the notion that our complicated immune systems are no better than the immunity manifested by so-called lower animals, will we also concede that our brains are no better? Not an easy question to answer. For now, I'll reiterate my original arguments: (1) you don't have to have a brain to be highly intelligent; (2) all living intelligence has the same basic architecture, an architecture that embodies, in some form, basic Darwinian principles; and (3) intelligence is an emergent property arising from large groups.

Before we explore the architecture of living intelligence in detail, we need to reverse course and examine the intelligence of the single cell. And, in doing so, we'll write the final chapter for poor Tom.

3

THE
ENZYME
COMPUTER

*The little things of life are as interesting as
the large ones.*

—Henry David Thoreau

Tom's flu shot would protect him against the ravages of pneumonia, but it couldn't shield him from that most dreaded risk of smoking: cancer. He arose one morning and started coughing, as he had done every morning for the past forty years. However, this particular morning ritual proved disturbingly different. Instead of clear secretions on his wadded tissue paper, Tom discovered thick ropes of sputum laced with dark blood. Although he was not a physician, Tom had lived long enough to realize what this signified. He headed straight to his local emergency room, where a chest X ray confirmed his worst fears. Whereas a chest film done only three months earlier had been normal, at the apex of his right lung now lurked a large and ominous shadow. His physicians knew that only cancer could grow so large in so short a time. The clock now ticked down the remnants of Tom's long life, and his future could be gauged in months instead of years.

Since cancer is a cellular disease, to understand Tom's problem we must understand the inner workings of the living cell. Cancer arises when one of our untold billions of cells transforms into a tiny sociopath and ignores the body's community standards. To function as competent members of a complex multicellular colony, all cells must obey certain rules that define their behavior. A neuron obeys one set of rules, a pancreatic cell another, but the cancerous cell, like the true sociopath, sets its own course, usually to the detriment of the community at large. What agency defines the body's rules, and how do normal cells know to follow them? And, more important for Tom, how do cells defy these rules to become cancerous rogues? Only by understanding these questions can we begin to comprehend the disease about to claim Tom's life.

In this chapter we'll explore the intelligence of cells, using cancer as our guide. This may seem odd, since at first glance, cancer looks like an example of biology gone stupid. Yet stop and consider what's happening: Tom's cancer cells were once normal lung cells. These deviant cells didn't enter from the outside, like the bacterial invaders that caused his pneumonia. Instead, they arose from within his body. Cancer is a transformational process, the act of one cell, equipped with a unique set of behaviors, transforming into a new cell possessed of a wholly different personality. We tend to think of cells as little machines built for a specific purpose, akin to automobiles, but they are much more than this. After all, a Ford Taurus is a wonderfully complex machine with all sorts of technological features, but when I park a Taurus in my driveway at night, I don't expect to find a bulldozer there the next morning. A Taurus that could reconfigure itself into a bulldozer, even by mistake, would be no ordinary machine at all. Thus, the ability of any machine to redefine itself as an entirely different entity—a quiescent lung cell becoming a relentless cancer cell—implies a degree of self-awareness and intelligence.

There is a common belief (even among many people who should know better) that cancer cells are simply "broken" versions of normal cells. In this view, the process of malignant transformation (normal cells becoming cancer cells) is no more sophisticated than some genetic fuse blowing. Cancer cells are indeed broken, but the process of malignant transformation is less like a fuse blowing than like an elegant piece of software crashing.

Cancer cells do more than divide aggressively. Tom's miscreant cells would soon detach from the main tumor mass and tunnel through the normal confines of his lung, heading straight for the brain and adrenal glands. The satellite tumors would then build more blood vessels to supply their increased demand for nutrients. Finally, they would counter any attempts by Tom's body (or his doctors) to restrain them. Cancer cells may be "broken" as far as the body is concerned, but they're cunning, relentless beings with a deadly agenda. The process of malignant transformation goes to the heart of life's central mystery, the intelligence of the single cell.

At this point, it appears as though I'm headed in the wrong direction. My first chapter examined the intelligence of bacterial communities, and the second chapter explored the intelligence of multicellular immune systems. Now, in the third chapter, I've backtracked, from global communities and organ systems to life's smallest unit, the cell. It would seem more logical to start with a discussion of the cell and proceed to multicellular communes—but there's a method to my madness.

All living things require a high level of intelligence to survive. So far, I've been developing the thesis that all living intelligence is an emergent property of cooperative societies. I've already shown that societies of bacterial cells and lymphocytes unite to create a high degree of social intelligence. To apply the same reasoning to the individual cell, on the other hand, I must define the cell as a "society" as well—but a society of what? My hypothesis appears to have struck a wall, namely, the irreducible nature of the cell. But all is not lost. We can indeed consider the cell a society, and a very complicated and robust one at that. The simplest cells are societies of enzymes, or, more specifically, *societies of biochemical reactions catalyzed by enzymes.*

The idea that a society can be constructed out of molecules and chemical reactions sounds outlandish and counterintuitive, which is why I chose to introduce the concept of social intelligence using bacteria and lymphocytes as the initial models. Although we can't imagine playing golf with an E. coli or having dinner with a lymphocyte, we can still concede that these small creatures do, in their own fashion, form interactive societies. Had I started "at the beginning" and examined the intelligence of the cell first, I would have had to introduce two difficult concepts simultaneously: (1) that intelligence derives from societies, and (2) that cells are societies of biochemical reactions. I trust that I've made some headway in explaining the first concept. Let's see if I can do likewise with the second.

We will begin with a little basic chemistry. Well, a lot of basic chemistry, actually, but I must take the time to build the foundation for what's to come. Life is, after all, a chemical process, and to understand life we must understand its chemical basis. This detour will take us very far afield, but stay with me. We'll get back to cellular intelligence—and cancer—in due time.

A chemical reaction occurs when combinations of atoms or molecules join (or break apart) to form different combinations of atoms or molecules—for example, when hydrogen and oxygen join to make water. In general, chemical reactions are reversible: they can go in either direction. Hydrogen and oxygen make water; water can decompose back into oxygen and hydrogen. A reaction is said to be at equilibrium, or in a steady state, when the forward and reverse reactions occur at the same rate, thus canceling each other out.

When a sealed glass bottle is filled with oxygen and hydrogen, the gas molecules randomly roam the container, bumping into the walls and also into one another. Occasionally, two hydrogen atoms bump into an oxygen atom with enough force to create a single molecule of H_2O—water. Initially, this forward reaction predominates. Because there is no water in the container at time zero, the reverse reaction—water breaking apart to form oxygen and hydrogen—can't begin until sometime later. At this early stage, our container is said to be "far from equilibrium," or at a point where the forward and reverse reactions are greatly unbalanced.

As water fills the container, the pace of the reverse reaction increases, since the more water there is in the container, the greater statistical likelihood that some will spontaneously break apart to form oxygen and hydrogen again. Furthermore, the forward reaction will slow down as oxygen and hydrogen are depleted. As fewer gas molecules are left within the container, the probability that they will bump into each other to form water declines and, eventually, the two reaction rates become identical: the net production of water (and the net depletion of hydrogen and oxygen) becomes zero. At this point, our sealed bottle reaches a *steady-state equilibrium;* the concentrations of the three chemical species swimming within—oxygen, hydrogen, and water—become fixed. That doesn't mean that the concentrations become equal. Under most conditions, the equilibrium state of this particular reaction dictates that nearly all the hydrogen and oxygen will eventually become bound up in liquid water, with only minuscule amounts of free hydrogen and oxygen remaining.

In this example, hydrogen and oxygen are called *reactants* and water is the *product.* The final concentrations of reactants and products at equilibrium are determined by the nature of the reaction itself. In the water reaction, the equilibrium concentration of the product (water) vastly exceeds that of the reactant gases. Hydrogen and oxygen prefer to coexist as water rather than remain as free, independent gases.

The fact that water is the preferred state does not mean that our container of hydrogen and oxygen will immediately condense into water before our very eyes. The rate at which a closed system reaches its unique steady state can be quite slow, even when the system starts out far from equilibrium. At room temperature, our bottle may not reach equilibrium for centuries. Thus, even though the equilibrium state is preferred, reaching that preferred state may take an extraordinarily long time. If I punch a tiny hole in a giant swimming pool, the pool will eventually drain completely—being "empty" is the preferred state of a lacerated pool. However, if the leak is tiny, months may pass before this preferred state is achieved.

Life depends upon its exquisite ability to manipulate the rate at which chemical reactions reach equilibrium. At the temperatures comfortable for living things, most chemical reactions, left to their own devices, reach equilibrium at a snail's pace. Without some means of speeding things up, life, as we know it, couldn't exist. This doesn't mean that life likes every reaction to achieve equilibrium. On the contrary, living systems would abhor seeing every one of their reactions at equilibrium, and for good reason: systems reaching such a total state of equilibrium are inert, dead.

Nevertheless, living systems spur reactions to equilibrium quickly. The two goals of avoiding complete equilibrium yet driving reactions to equilibrium at great speed seem grossly incompatible. The solution to this paradox can be found in the fact that life is an open system. The sealed bottle of hydrogen and oxygen was a closed system: nothing could get in or out once the reactions began. In an open system, reactants and products freely enter and leave the system, thereby preventing it from ever reaching equilibrium.

To see why open systems can't fully equilibrate, let's reconfigure our oxygen/hydrogen bottle as an open system; oxygen and hydrogen are now pumped in continuously and the resultant water vapor is removed. This system can't reach equilibrium, because the reverse reaction (water breaking down into oxygen and hydrogen) will never occur. Water vapor constantly escapes the container and prevents the reverse reaction from ever getting started. Let's drive the system to equilibrium faster by igniting the incoming gas mixture with a match. We now have a hydrogen furnace. Within the furnace, the chemical reaction ferociously chases equilibrium, but the continuous flux of gas into the furnace—and egress of water vapor out—so heavily favors the forward reaction that the furnace will never reach a true steady state. Like a feverish mule lurching after a carrot on a stick, the flame runs toward equilibrium but can never reach it. Some of life's metabolic pathways act like open systems, drawing in food and oxygen and emitting carbon dioxide and wastes. Fire is the second-best example of an open system rapidly seeking, yet never achieving, equilibrium. The best example is life. From a chemist's perspective, the metabolic furnace of life is simply a room-temperature flame.

Heat is the simplest way of driving most systems to equilibrium quickly. The temperature of matter is defined by the average kinetic energy of the atoms and molecules comprising it. Heating a mixture of oxygen and hydrogen accelerates the gas atoms and increases their chances of randomly colliding to form water molecules. Living systems can't exploit this trick, because they contain many organic molecules that can exist in narrow temperature ranges. Life must rely on another trick to accelerate reactions, something as good as heat: protein catalysts known as *enzymes*.

A catalyst is any substance that accelerates a chemical reaction toward equilibrium. In general, catalysts act by increasing the probability that chemical reactants will find each other. As has been seen, heat works by globally increasing the average motion of atoms and molecules. Just as people running randomly through a mall at high speed are more likely to crash forcefully into each other than people walking slowly, thermally agitated molecules and atoms are more likely to make contact with enough energy to produce a chemical reaction.

(Flat surfaces can also act as catalysts by binding reactants, thereby limiting their mobility to two dimensions instead of three. The restricted freedom of reactants on surfaces also increases their likelihood of colliding to form products. Life's earliest reactions, at a time before the advent of true enzymes, may have been partially catalyzed by being absorbed onto clay particles in the primordial oceans.)

Today, life relies almost exclusively on enzyme catalysts (although "surface catalysis" may still take place on two-dimensional cellular membranes). Enzymes accelerate chemical reactions by bringing two or more reactants into close proximity. Like antibodies, enzymes are equipped with customized binding sites designed to grab reactants and hold them together in a way that facilitates chemical bonding. An enzyme can also bind a single molecule and break it into pieces. Enzymes function like carpenter's clamps, holding reactants in the correct orientation so that they may be glued together or sawed apart in the appropriate fashion.

A better analogy for enzyme behavior may be that of a dating service. In a world full of people, it could take years for two people with identical interests to meet at random. Dating services use intelligent data analysis, such as computer matching, to accelerate the identification of compatible partners. Likewise, enzymes use the intelligence built into their binding sites to match compatible molecules much more quickly than would be possible if they were left to seek each other out randomly. Chemically, enzymes serve a function similar to heat: they drive reactions to equilibrium by increasing the likelihood that reactants will interact within a given period of time. The enzyme's intelligence replaces thermal energy by altering the statistical behavior of large populations of molecules and atoms.

We should stop and ponder this point. Earlier I referred to life as a fire burning at room temperature. The heat of life's cold flame resides in the binding sites of enzymes. The binding site (more specifically, the informational content of the binding site) becomes a thermodynamic entity, like temperature or entropy. We now tread into the difficult terrain first explored by the great Scottish physicist James Clerk Maxwell. Over a century ago, Maxwell proposed a form of perpetual motion machine built around an atom-sized creature now known as "Maxwell's demon."

The temperature of matter is a measure of the *average* velocity of the molecules comprising it. In any large collection of molecules, like a room full of air, some molecules move slower than average and others move faster. The distribution of atomic velocities follows a modified "bell curve" known as the Boltzmann distribution. Maxwell concocted a simple thought experiment: Take a room of air at a certain temperature (T). The room is totally sealed save for a tiny door, which is operated by the demon. The demon is intelligent; he can tell which molecules move faster or slower than average. When a fast molecule heads for the tiny door, he opens it and allows the speedy molecule to escape the room. Conversely, when a slow molecule approaches the door, he closes it, reflecting it back into the room. Over time, the demon will selectively lower the average velocity of the remaining molecules, thereby cooling the room. He expends no energy, since opening and closing the door can be made extremely easy. Thus, the demon becomes an air conditioner that requires no energy to operate. He generates work without expending energy, acting as a true perpetual motion device. As anyone who pays the electric bill for a real air conditioner knows, this seems impossible. And it is. Cooling a room without expending energy is a violation of the Second Law of Thermodynamics. So what's happening?

The paradox of Maxwell's demon wasn't solved until the 1950s, when theorists showed that the demon's act of measuring molecular velocity and deciding when to open the door indeed required energy—informational energy. The demon's intelligence comes with an energetic price tag after all; acquiring, processing, and acting upon information requires work. Thought, even at the molecular level, requires effort, and effort requires energy. The twentieth century saw the rise of information theory and the realization that information is not an abstraction, but a physical entity with tangible thermodynamic consequences.

The impact of information on thermodynamics is quite tiny, far too negligible to have any effect on macroscopic systems like brains and computers. At the atomic level, however, the effect of informational input—Maxwell's demon—could impact physical processes in such a way that they deviate from classical thermodynamics. The strange brew of chemical reactions known as the cell's cytoplasm may represent the nonclassical juxtaposition of chemical and informational thermodynamics.

Enzymes give us our first glimpse of the potential interplay between information and classical chemistry in living systems. They act like protein versions of Maxwell's demons, altering the thermodynamics of chemical reactions, manipulating molecules, and skewing the statistical likelihood that chemical reactants will find each other in a random chemical mixture. An enzyme's binding site mimics the demon's little doorway, opening for the correct reactants and closing for the wrong ones. Enzymes become the transistors of the cellular computer.

Before returning to the subject of cell biology and, in turn, to the subject of cancer, we need to take one additional detour: we must reconsider the concept of nonlinear behavior in greater detail.

To illustrate the difference between linear and nonlinear behavior, compare a radio's volume knob with a common light switch. Both are "input-output," or I/O, devices, in that they convert a certain input into a corresponding output. The volume knob converts the twisting motion of my hand (input) into a change in sound volume (output); the switch converts the up-and-down motion of my hand (input) into an on-off effect on a distant light bulb (output).

The volume knob is a linear I/O device because a plot of knob motion versus sound volume will be, more or less, a straight line. However, a plot of a light switch's motion versus the intensity of light will not be a straight line. As I move the switch upward, the bulb remains off until I cross a certain threshold, at which point the light pops on. The light switch is *bi-stable*—it has two steady states, on and off. Go to the nearest light switch and try to balance it midway between these two states—not easy, is it? The switch wants to pop into one of its two stable states; it dislikes lingering halfway in-between.

We can push the switch into either state at our whim and it will stay there until we switch it back again. The bi-stable switch becomes a sort of memory device. We can use a series of switches to store coded information indefinitely. The word *stable* here is relative; the states are stable until some external agency decides to change them.

To build any intelligent machine, we must have components that are, at minimum, bi-stable. Why? Because the minimum encoding language for information processing is the binary system, comprised simply of 1s and 0s. To store binary data, we need things that can store two separate states, on and off, 1 and 0, up and down, and that can be reversibly switched between states by some external force. Suppose we have ten billion electronic devices lined up in a row, all possessing just a single steady state. These devices form no pattern and carry no information, because they're all the same. But make them bi-stable, equating the two steady states with the dot and the dash of Morse code, and look again. We can now go down the row and flip some devices into the dash position and some into the dot position. With this simple system, we can write Shakespeare, or the Bible, or anything we want, using the Morse alphabet. (As will be discussed later, some forms of computation go beyond simple bi-stability by utilizing a continuous range of values between 1 and 0, but for now, the simple bi-stable switch seems to be the least we need for a practical computational machine.)

Not surprisingly, transistors are bi-stable devices. So are neurons. These devices are either "on" or "off" and can be switched between the two polar states by external forces. In this way, complex information can be encoded in large groups of transistors and nerve cells. Building informational machines out of transistors and nerve cells is comparatively easy. If the cell itself is an intelligent machine, it must also be assembled from (at least) bi-stable units capable of encoding binary information. Cells, as we know, are made from chemical reactions, no argument there, but once again, my thesis (the cell is intelligent) hits a wall. It seems that we can't build an intelligent machine out of ordinary chemical reactions, because most chemical reactions are mono-stable under conditions suitable for life. They typically have one, and only one, stable state. Thus, such chemical reactions alone seem incapable of serving as the cell's bi-stable switches.

Well, this isn't strictly true: there are some inorganic chemical reactions that are bi-stable. One famous example is the iodate-arsenous reaction:

$$IO_3^- + 3H_3AsO_3 \rightleftharpoons I^- + 3H_3AsO_4$$

which, depending on the initial conditions, will gravitate toward one of two possible stable states, a state producing a large amount of I^- and one producing little I^-. The chemist can manipulate this reaction like a switch, popping the reaction into either state at will. Theoretically, a chemical computer could be built using bi-stable iodate-arsenous reactions in place of bi-stable transistor switches. For example, we could line up a series of vats filled with the iodate-arsenous reaction and use the high and low concentrations of I^-, in place of dots and dashes. With enough vats, we could store vast amounts of information. The vats needn't be very big; after all, chemical reactions occur within the confines of a single bacterial cell and smaller spaces—a chemical computer could be made very tiny indeed. Allen Hjelmfelt and his coworkers in Wurzburg, Germany, proposed just such a device in a *Science* article written in 1993, although to my knowledge it has never been built.

Although we don't know everything about living cells, we're fairly certain they're not made of bi-stable iodate-arsenous reactions; we also know that most common chemical reactions aren't bi-stable. So what's the answer? How can the chemical reactions making up cells—reactions with only a single steady state—be linked to form an information-handling device? The answer, once again, can be found in enzymes.

In general, for a chemical reaction to exhibit bi-stability, some catalyst must be present. In the case of inorganic reactions, the catalyst is either one of the reaction products themselves or a two-dimensional surface (surface catalysis). When one of the reaction products catalyzes its own production, the reaction is said to be *autocatalyzed*. Autocatalysis injects a feedback loop into the reaction, creating nonlinear behavior. The iodate-arsenous reaction and the equally famous Belousov-Zhabotinski (BZ) reaction exhibit autocatalysis and manifest nonlinear behaviors, including bi-stability. A number of ordinary chemical reactions can undergo autocatalysis during high-temperature combustion, but under conditions compatible with life, most reactions do not. For its computational machinery, life requires more bi-stability than simple inorganic chemistry can readily supply. To create bi-stable reactions on demand, life uses enzyme catalysts.

In 1993, the same year that Hjelmfelt and his colleagues proposed building a molecular computer, Robert Jackson of Agouron

Pharmaceuticals in San Diego published an article in the *Journal of the National Cancer Institute* entitled "The kinetic properties of switch antimetabolites," in which Jackson showed that any reaction can have two or more steady states if catalyzed by an enzyme. His idea wasn't entirely original—J. Reich and E. E. Sel'kov made a similar prediction decades earlier—but Jackson presents the idea of enzyme-induced bi-stability with particular clarity.

Enzymes found in living cells are subject to feedback control. In addition to having a binding site that catalyzes a given chemical reaction, enzymes have regulatory sites that can turn their enzymatic function on and off. These regulatory sites are, in turn, controlled by other molecules, often the same molecules involved in the catalyzed reaction. When a reaction product activates its own enzyme, autocatalysis occurs and the reaction converts to a bi-stable switch, but this isn't the only way enzymes can generate bi-stable behavior. Just as enzymes can be turned *on* by their own products (autocatalysis), they can also be turned *off* by the products of other reactions (cross-inhibition). Like autocatalysis, cross-inhibition introduces feedback loops that generate bi-stable behavior. Using computer modeling of enzyme autocatalysis and cross-inhibition, Jackson nicely demonstrated how feedback loops induce bi-stability in enzyme-catalyzed reactions. In fact, an enzyme-catalyzed reaction can manifest multiple stable states even when it is quite far from its true equilibrium condition. In other words, enzymes introduce "quasi-equilibrium" states even in open reactions that, theoretically, can never reach any true chemical equilibrium.

Consider the case of cross-inhibition. Without going into the gory mathematical details, I'll just say that feedback forces two reactions to behave like a bi-stable "flip-flop" circuit inside a desktop computer. A flip-flop consists of two transistor amplifiers wired together so that the output of one inhibits the output of the other.

This is identical to the cross-inhibition of enzymes, where one reaction inhibits the rate of another reaction somewhere else. When one amplifier is at maximum output, its partner is driven to its minimum output. The flip-flop has two stable states: first amplifier on, second off; first amplifier off, second on. An external force can flip or flop the device between two states like a switch going on and off. Cross-inhibition mimics a playground seesaw, where two counterbalancing weights compete to keep the system in one of its two stable states. In the case of two cross-inhibiting reactions, one will proceed quickly and the other slowly. This situation can be flip-flopped—the fast reaction slowed and the slow reaction accelerated—as the need arises.

In an autocatalytic system, a product helps catalyze its own production, creating a seesaw effect within a single reaction. When the reaction proceeds slowly, the lack of product tends to keep the reaction slow; if the reaction proceeds quickly, the abundance of product tends to keep it fast. Thus, the reaction works to keep itself in its present state (whether slow or fast) and resists change. In this way, autocatalysis creates two stable reaction states: very slow and very fast. Left to its own devices, the reaction will remain slow or fast until some outside agency (1) removes enough product to trigger the slow state, or (2) adds sufficient product to switch it to a rapid state. In both autocatalysis and cross-inhibition, the reaction stays in a stable state until an outside force switches it; that's exactly the type of bi-stable device a computational system needs. If I'm building a computer, I want switches that I can manipulate and that will hold their positions until I move them again.

Enzymatic reactions coupled to each other or to themselves via feedback loops are like the coupled amplifiers in a computer flip-flop. With the advent of enzymes, cells acquired the bi-stable components to assemble a computational machine. Although the occasional inorganic reaction manifests enough autocatalysis to become bi-stable all by itself, a properly designed enzyme can make any reaction bi-stable, or even multi-stable.

Earlier, I said that life didn't like equilibrium. So why would living things use equilibrium states like so many 1s and 0s in a computational machine? Because equilibrium states still govern the behavior of a system even when it's "open" and far from reaching any equilibrium. A flame never reaches equilibrium, but its behavior is still dictated by the pursuit of an elusive equilibrium state. Consider, for example, that my "equilibrium point" with regard to gravity resides at the center of the earth. Left to reach equilibrium, I won't stop falling until I reach that point. Technically, as I sit here typing, I'm far away from gravitational equilibrium—four thousand miles away, to be precise. Nevertheless, gravity still governs my dynamics. Everything I do is dictated by the inexorable pull toward my gravitational equilibrium state, even though I will never (I hope) make any journey to earth's core. Likewise, all reactions, even those far from equilibrium, must still obey the pull of their equilibrium states.

Dr. Jackson was interested in cancer, not enzyme computers. After demonstrating that any cellular reaction can have two or more stable states, he argued that cancer cells were simply normal cells that had one or more of their constituent reactions trapped in the wrong stable states, hence his use of the term *switch pharmaceuticals*. Instead of using drugs to kill cancer cells, Jackson proposed using them to switch faulty reactions back to their correct stable states, thus switching faulty tumor cells back into normal cells without the need for killing them. But Jackson's work is even more important than this; it also gives us a clue as to how intelligent molecular-sized machines can be built from chemical reactions.

To summarize, so far we know these four things:

1. living cells are built from chemical reactions;
2. living reactions rely on enzyme catalysts to provide the "cold flame" needed to drive reactions quickly toward equilibrium at low temperatures;
3. an intelligent machine must be assembled using nonlinear switches possessing at least two stable states; and
4. under the conditions suitable for life, most simple chemical reactions (save for a limited class of inorganic reactions) have only one steady state and can't be used to encode binary information.

We can now add the one fact that breaks us through the wall:

5. protein enzymes can also insert nonlinear feedback loops into virtually any reaction, converting it to a bi-stable (or multi-stable) switch subject to external manipulation.

Conventional wisdom states that molecular intelligence began with the advent of nucleic acids, but I believe the true origin of organic intelligence can be traced to the evolution of enzymes. Enzymes imbue chemical reactions with the nonlinear properties essential for assembling themselves into computational engines. In effect, life simulates naturally bi-stable reactions and exploits their oddball properties to make the intelligent little enzyme-clockworks we now call cells. (The conventional wisdom still wins, in a sense. The first enzymes were probably made of RNA, not proteins. My point holds regardless of the chemical nature of the first enzymes.) It's possible that some of the first living reactions were like the inorganic iodate-arsenous and BZ reactions and possessed a certain degree of non-enzymatic autocatalysis, or, as I mentioned earlier, the first catalysts could have been two-dimensional surfaces. Regardless of how it first appeared, bi-stability became the foundation of the intelligent biological machine. To understand molecular intelligence (and that's what life is, molecular intelligence), we must understand molecular bi-stability. To seek the origins of life, we must seek the origins of the chemical complexity that allows molecular computation.

Nucleic acids are great repositories of information, but they can't think. The *Encyclopaedia Britannica* contains a vast amount of information, but it has no intelligence. Intelligence is a dynamic process and, in the cell, the dynamism comes from the cold flame, the enzyme-driven reactions that unite to form an intelligent being on a molecular level.

Let's now consider the difference between mono-stable and bi-stable systems in a little more depth. Look at the following two diagrams:

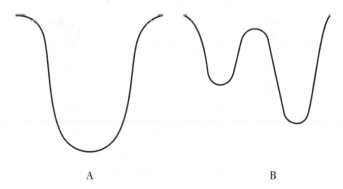

A   B

Let's assume that the diagrams are side views of roller-coaster tracks. When we place a coaster car on the steep slopes of either track, that car will be "far from equilibrium," assuming that the valleys at the bottom represent steady states of the coasters. How do the two coasters differ?

If we put a car on either the right or left slope of coaster A, the car will roll to the bottom and, after some to-and-fro oscillations, it will come to rest at the lowest point of the track, a point representing the only steady state for this system. This coaster, possessing a single steady state, is *mono-stable*. The final state of coaster B, on the other hand, clearly depends on where we place the car initially. If we start the car from the right, near the bottom, it will slowly roll into the valley on that side and stay there; the reverse will happen if we place it on the other side. If we start a car higher on either slope, it may speed down so quickly that it crosses the hump in the middle before bouncing back and forth between the two valleys. Predicting exactly in which valley a fast car will end up will be difficult. Finally, if we perch the car precariously on the middle hump, it will stay there briefly until some slight jar, such as a gust of wind, starts it rolling in either direction. Like the light switch, coaster B won't hover long between two stable states. Coaster B is bi-stable; which of the two stable states the system comes to rest in will depend exquisitely upon where we *initialize* the coaster car—where we place it initially on the track. The output of coaster A, in contrast, is independent of where we initialize the car.

Both A and B show nonlinear behavior, but only B forms a meaningful input-output device. If we define "input" as where we initialize a coaster car and "output" as where the car comes to rest, we can conduct an experiment with each coaster by randomly placing a car of a certain weight at various places on the track and recording where the car ends up. For coaster A, this is a trivial exercise—all outputs are the same. For coaster B, however, we'll find that some starting points end up in the left stable state, some in the right. The precise behavior of coaster B will be influenced by things like how heavy the coaster car is, how slippery the track, and so on. This sort of subtle bi-stable behavior isn't all that different from the behavior of transistors and nerve cells. Believe it or not, we could build a computer out of many coaster Bs if we had enough space. Coaster B is a usable I/O device, thanks to its bi-stable nature.

Like coaster A, mono-stable chemical reactions seek the same predictable equilibrium no matter what the initial conditions. Regardless of how we filled (initialized) our glass bottle with oxygen and hydrogen, the water reaction will seek its one steady state. Multi-stable reactions, on the other hand, will seek different stable states depending upon where we initialize them. Let's return to the iodate-arsenous reaction for a moment. This reaction resembles my coaster B in that it possesses not one, but two valleys. In the case of chemical reactions, the "coasters" that represent their dynamic behavior aren't in any tangible space defined by $x, y,$ and z coordinates, but in a hypothetical space in which the "dimensions" are defined by the concentrations of the products and reactants.

One valley of the iodate-arsenous coaster is a stable state producing a low concentration of free iodine product; the other valley represents an alternate stable state producing high concentrations of iodine product. By varying the concentrations of products and reactants at the onset, a chemist can initialize the iodate-arsenous "coaster car" anywhere on the track and observe into which valley the reaction finally comes to rest. Jackson showed that enzymatically catalyzed chemical reactions also tend to look like coaster B. In fact, such reactions can have equilibrium "landscapes" far more complicated than those depicted in the simple diagrams above. Nothing prevents us from making a coaster with three or more valleys. Moreover, my coasters have only two dimensions, defining two variables (up and down, right and left), while real physical systems, including chemical reactions, often have many more variables. We could, for example, define a hypothetical chemical reaction like:

$$A + B + C \rightleftharpoons D$$

where we have four distinct variables, namely, the concentrations of molecules A, B, C, and D. The equilibrium landscape of this reaction—the various stable states plotted as a function of values of these four variables—becomes a four-dimensional hyperplane. Steady states now become dimples in this plane, with the reaction rolling like a ball in search of the nearest steady-state "dimple." Hyperplanes? Things are getting very complicated indeed.

And that's good. To build intelligent machines, we need such complicated behavior. Conventional inorganic chemistry, with its tedious plodding to equilibrium and mono-stable reactions, won't do the job, but thanks to enzymes, even so-called simple chemical reactions aren't so simple anymore. Inside the cell's enzyme stew, organic chemical reactions become sophisticated I/O devices capable of switching between different stable states in response to environmental factors. Moreover, the products of one reaction are invariably the reactants in some other reaction, such that all reactions inside the living cell become coupled into one massive chemical web, or (more precisely) a huge chemical network.

It's time to introduce the formal concept of a network. The network concept will begin to dominate our discussion, for reasons that will become apparent in the ensuing chapters. Up until now, I've often used the less formal term *society* for *network*. For the purposes of this book, *a network is a collection of bi-stable (or multi-stable) I/O devices connected together, with the output of one I/O device serving as input for others in the same network*. Networks have very interesting properties, as you will learn in this and the next two chapters. The brain is a network, but, curiously enough, the personal computer I'm typing on now is not. Conventional computers don't fulfill all the requirements of a true network.

In a cell, the "output" of any chemical reaction becomes the "input" of one or more other reactions. The product of one reaction is invariably the reactant in another reaction elsewhere. Moreover, the enzyme catalyzing one reaction may be activated or inhibited by products and reactants of other reactions. The cell's multitude of reactions are hopelessly intertwined in a true network. If we accept the idea that the enzyme-catalyzed reactions can act as multi-stable I/O devices, we understand how the cell becomes a network of connected reactions, thus forming an enzyme computer similar in principle—though infinitely superior in power—to the iodate-arsenous computer of Hjelmfelt and his colleagues. Just as the brain is a network of connected neurons and the immune system a network of connected lymphocytes, the cell is a network of chemical reactions.

(We are now getting our first inkling that all forms of intelligence, alive or otherwise, share common features. All networks, whether built from transistors, molecules, or nerve cells, obey certain universal rules.)

If we consider that the number of chemical "neurons" inside the smallest cells exceeds the number of neurons in the biggest brains, we arrive at the unsettling notion that a cell may approximate our own intelligence, at least within its own peculiar environment. Before scoffing at this notion, consider a recent experiment demonstrating that slime molds can calculate the shortest path through a maze when searching for food—a remarkable feat of computation for such a "primitive" organism. Recall also that our exalted brain, indeed the entire human body, springs from a solitary fertilized ovum, a single-celled engineer invisible to the naked eye. We pat ourselves on our collective back for erecting comparatively crude structures like Notre Dame cathedral. How pathetic such brain-built constructs look compared to the accomplishments of an ovum!

One final digression before returning to cancer and poor dying Tom. I spoke earlier of an equilibrium landscape, a crucial concept that needs to be fleshed out further. An equilibrium landscape is a graphic representation of the behavior of some dynamic system; the steady states become valleys, or basins, in this landscape. My coaster diagrams were very simple examples of such landscapes.

The word *landscape* evokes a bucolic meadow, with gentle hills and sloping valleys. In reality, equilibrium landscapes can be any number of dimensions, but because few mortals can think in dimensions exceeding three, I'll stick with the meadow metaphor. As will be seen in the next chapter, large networks are described as extraordinarily complicated "meadows" with countless hills and valleys of different sizes and shapes.

Once again, keep in mind that these landscapes exist in virtual spaces where dimensions can be defined in any way we (or a network) choose. In chemical systems, the dimensions are defined by the concentrations of the different molecular species involved. We could define one dimension as temperature, another as the frequency of a certain sound wave, yet another as wind velocity—it all depends on what dynamic system we wished to model.

The landscape portrays all possible dynamic states of a network in graphic form, and the valleys, as has already been seen, represent steady states. (Another common term for a valley in an equilibrium landscape is *attractor basin,* or simply *attractor*—so named because the system is always being "attracted" to one of its steady states.) When we build any system from scratch, whether that system is a simple chemical reaction like our hydrogen/oxygen bottle or a complex network like the human brain, we must define its initial parameters. For example, in our closed bottle, we define how much oxygen, hydrogen, and water are in the bottle initially. When we initialize the system in this fashion, we are, in effect, dropping it onto a specific geometric point within its equilibrium landscape, as if we were dropping a wagon from the sky onto a predetermined point in the meadow. Once initialized, the system, like the wagon, begins rolling toward the nearest valley via the shortest possible path.

The equilibrium landscape for the water reaction is an idealized depiction of the behavior of all possible combinations of oxygen, hydrogen, and water. When we initialize the system, we are picking a point on that idealized landscape corresponding to a real system at a real point in time. That real system must then follow the geography of the idealized landscape as it seeks its final resting state.

Needless to say, there is no gravity in our abstract meadows, so what drives our hypothetical wagons into the attractor basins? The same thing that makes wagons roll down hills: the need to achieve the lowest possible energy state. The laws of thermodynamics require that all physical systems seek their lowest state of energy. A more technical, but less intuitive, name for an equilibrium meadow is "E surface," short for energy surface. *The attractor basins are those parts of the E surface where the system in question is at its lowest energy.* Coaster cars have the least amount of stored energy at the bottom of a hill; the hydrogen-oxygen reaction has the least stored energy at equilibrium. Conversely, coaster cars have their greatest amount of stored energy at the top of a hill, and chemical reactions have their greatest stored energy when far from equilibrium. As a general principle, we can dispense with the idea of gravity and define motion within any equilibrium meadow—any E surface—as the act of traveling from a point of higher energy to a point of lower energy via the quickest possible route. Strangely enough, all network computation can be reduced to an exercise in geometry. (We'll revisit the philosophical implications of this in the final chapter.)

Networks built out of many connected I/O sub-units can be configured in such a way that the E surface becomes an abstract representation of some problem to be solved; the valleys, or attractors, then represent various solutions to that problem, some good, some not so good. (Trust me on this. The details will come later.) By linking, or mapping, a network's internal equilibrium landscape to some external problem, one can convert the network into an intelligent machine. Computer scientists recognized this fact about three decades ago; life, on the other hand, recognized it eons earlier.

How does one problem have so many different solutions? Consider this dilemma: where should someone invest $100,000? As any financial advisor knows, that all depends on the person. A young person can take more risks with his or her money than an older person nearing retirement. It also depends on what markets are producing the best yield at any given time, and over what period the investment will be made. Some markets offer good short-term yields; others render better returns in the long term. The solution to this problem depends on how we initialize it, or how we define our variables.

This problem can be "mapped" onto an equilibrium landscape in an abstract space defined by various financial and demographic variables. This landscape represents all possible solutions for all possible investors. When we set the initial parameters more specifically—the age of the investor, the level of risk he or she wants to take, the behavior of the markets at a certain point in time—we initialize the system and start it rolling toward the nearest attractor representing the nearest (and, one hopes, best) solution for that investor. Of course, the system could roll into a poor solution, too, resulting in the investor's bankruptcy. As a network learns, it figures out how to reconfigure its E surface so that the best solutions remain and the worst solutions are eliminated.

If this all sounds very theoretical, guess again. This morning I logged onto the Internet and found several companies marketing "neural network" software that will help me manage my investments using precisely this method, namely, by creating a network with a landscape filled with attractor solutions to my investment needs.

This software, which employs simulated I/O switches ("neurons") linked into a network, is called *neural network* software because it uses an approach believed to have originated in the brain (brain chauvinism again). Indeed, brains do use this approach, as will be covered in depth in a later chapter, but they didn't invent it. The whole point of the present chapter is that the network/landscape paradigm of problem solving, a paradigm used by brains and neural network software, originated at the cellular level at the very dawn of life.

While we're on the subject of brains, let's digress briefly and consider a quick example of a brain landscape. My memory for 1960s pop tunes is really an E surface landscape created by my network of neurons. The landscape exists in a space defined by the tonal qualities and sequencing of notes within songs. Each song is a valley, or attractor, in this vast meadow of remembered tunes. The moment I hear one note of a song, a wagon drops onto the meadow and starts rolling in my head. The wagon isn't plopped down just anywhere, however. Based on that one note, the wagon is initialized in that area of the landscape containing all songs beginning with that note. With each successive note, the wagon's course is adjusted until it falls into an attractor valley. It may not be the correct valley, but it will fall into a valley nevertheless.

If I know the song very well, the valley corresponding to that song will have very steep walls and the wagon will rapidly gravitate to the bottom of the correct valley. In fact, I may not need any more than two or three notes to recognize and remember an entire song. The wagon will roll into the valley after only a brief sojourn on the landscape, and I will identify the song precisely. If, on the other hand, I barely recall the song, the valley will be very shallow and the car may roll past the right valley and fall into an adjacent (wrong) valley, where it will be trapped. If limited to hearing only one or two notes, I may be convinced it's another song entirely. Landscapes aren't foolproof—a landscape error is about to kill poor Tom.

My song landscape is created through hearing and learning more and more songs. Learning, as it applies to networks, becomes equivalent to carving deeper and steeper valleys into their unique landscapes. Thus, landscapes can be used both for problem solving and for memory storage. In truth, what passes for problem solving in living things may be little more than elaborate feats of landscape memory.

In the beginning of this book, I discussed the concept of pattern recognition, the ability of living things to identify a correct pattern given incomplete information. Pattern completion comes easily to networks, because they employ the landscape approach to problem solving. If the bottom of an attractor valley represents the perfect solution to a problem, a network doesn't have to be initialized with complete information to reach that perfect solution. In other words, we don't have to dump the wagon right at the bottom of a valley for it to end up there; we have only to dump the wagon close to the valley and it will roll there on its own. If the imperfect information provided proves sufficient to initialize the network near the right attractor, the network will come up with the true solution by itself, filling in the gaps as it rolls down the hill. A few notes of a pop tune are sufficient to initialize my brain near the correct tune; once in the tune's valley, my brain can complete the missing notes. Pattern completion is an emergent property of the E-surface approach to memory and problem solving.

Clearly, the landscape represents all that the network "knows" at a given time. For intelligent networks, the act of learning continuously modifies the landscape's terrain, creating or altering attractor valleys in accordance with past experiences. At inception, an ignorant network possesses a flat landscape. As the network gains experience, the landscape heaves and dips to store that experience in its many attractor basins. A living informational landscape is typically dynamic, changing over time in response to the network's learning. In some cases—such as the behavioral instincts wired into the brain—a portion of the network's topography is programmed in advance and thus not subject to further modification. We'll revisit this point again very soon.

But how can experience be translated into geometric fluctuations of an E surface? After all, the equilibrium landscapes of things like the hydrogen-oxygen reaction aren't learned; they're fixed forever by nature, by the chemistry of the reaction itself. As it turns out, networks have a built-in mechanism for changing their E surfaces. But this difficult concept must also wait for the next chapter, when we deal with internal network weighting and connectivity.

◎

What does all this talk of steady states, attractors, and networks have to do with cancer cells? The answer lurks in a single word: Dolly.

Dolly was a sheep, the first mammal to be cloned. Cloning is the act of extracting the nucleus—the central sac of chromosomal DNA that determines the cell's identity—from the cell of one animal, inserting it into the ovum of another, and implanting the hybrid egg into a receptive uterus. If all goes according to plan, the ovum will spawn a new animal genetically indistinguishable from the one donating the nucleus. Just about any cell in the body can be used to clone a whole animal—a skin cell, a muscle cell, it doesn't matter. Any cell, that is, except eggs and sperm (cells known as gametes). Gametes carry only half the chromosomes of somatic, or non-gamete, cells, for the obvious reason that one egg and one sperm unite to give the full complement of chromosomes at fertilization. Moreover, gametes carry scrambled versions of the host's genes, so that the host's offspring will have genetic variability. Thus, one of my sperm cells could not be used to clone an identical copy of me, although (ironically enough) a cell from one of my hair follicles could.

Cloning illustrates an amazing fact, namely, that all non-gamete cells share an identical chromosomal makeup. Every cell in my body, save for my sperm, contains the full blueprint of the complete me. Thus, a neuron and a muscle cell from one body are, in a genetic sense, identical twins, the very same cell. But a neuron doesn't look or act anything like a muscle cell, so how can this be?

During embryogenesis, the fertilized egg divides repeatedly and its progeny begin acquiring the unique characteristics that will mark their future role in the mature organism. The earliest cells are called *pluripotential cells*, because they will eventually yield many different types of specialized tissues. The process by which pluripotential cells diverge and become nerve cells, bone cells, etc., is called *differentiation*. In our own bodies, all cells exist in a state of full differentiation, save for certain cells in the bone marrow, called stem cells, that differentiate into mature blood cells throughout our lifetimes.

Because the body's differentiated tissues vary so drastically, early biologists assumed that the act of differentiation must be genetically driven. They believed that differentiation mimicked geological evolution and that cells achieved their adult configuration via mutation, chromosome deletion, or some other permanent genetic modification. Bolstering this idea is the fact that differentiated tissues divide to produce other tissues just like themselves; when a pancreatic glandular cell divides, it produces two pancreatic glandular cells. In this paradigm, each somatic tissue type represents a genetically distinct lineage, a unique species of cell.

But cloning proved this paradigm wrong, for if different kinds of cells arose through genetic mutation, then differentiation should be irreversible. Yet we now know that it isn't: the nuclei of pancreatic cells can be cloned back into an entire organism. So too can the nuclei of pituitary cells, muscle cells, or any other somatic cells. Blood cells can even be removed from an adult's bone marrow and tricked into becoming liver cells. These findings prove beyond all doubt that the mammoth differences between a muscle cell and a neuron are neither genetic nor permanent. Muscle cells and nerve cells are, quite literally, simply the same cell, each temporarily portraying a different role in that elaborate stage production known as the body. The chromosomal makeup of a cell defines its *genotype,* while the actual appearance and function of a cell defines its *phenotype.* A cardiac muscle cell and a nerve cell taken from my body are of one genotype, but express very different phenotypes. Differentiation is a quantum thing. In the mature organism, a cell can be a muscle cell or a bone cell, but not something in between. Differentiation has no "shades of gray." Moreover, once a cell commits to being a muscle cell, it usually stays a muscle cell forever. Although differentiation is not irreversible, a cell can't change phenotypes easily.

This behavior sounds suspiciously like our roller coaster B. Cells prefer to land in stable steady states known as phenotypes; they aren't welded into those states, but they can't be easily budged out of a valley without considerable force being exerted upon them. In our coaster example, a car can be in different valleys and still be the same car. Likewise, a cell can be in different phenotypic states and still be the same cell. The intrinsic identity of the cell—the genotype—remains the same, regardless of what phenotypic valley the cell inhabits at any given time. Cells, again like coaster cars, prefer not to linger on the unstable "humps" separating differentiated valleys. Once a cell is in a valley, it will stay there (along with its progeny). As will later be discussed in more detail, the landscape approach explains not only the stability of cellular phenotypes, but the stability of multicellular species as well.

The approximately one hundred phenotypes within my body (muscle, nerve, bone, and so on) are merely different steady states of my one unique genotype. The enzyme computers within my cells, as defined by the cytoplasmic chemical networks, come equipped with a predetermined *phenotypic landscape* in which the attractors are the various differentiated phenotypes comprising my body. Different genotypes provide for different phenotypic landscapes—for example, the genotype of a jellyfish will program for a simpler landscape containing the attractors needed to define the less diverse number of jellyfish tissues.

During embryogenesis, newly emerging cells are initialized near different attractors in the landscape—for example, cells initialized near a muscle attractor fall into a steady-state muscle phenotype. The miraculous process of assembling a body becomes akin to tossing a million marbles onto an undulating tabletop and letting them roll into the appropriate basins. But what defines the shape of the tabletop, the E surface for differentiation? Genes, obviously—I'm human because my genes say I am—but that can't be the whole story. Because of brain chauvinism, we believe that all complex action derives from some centralized agency, yet we now know that a cell's DNA doesn't dictate what form that cell will take. The vast differences between a nerve cell and a muscle cell lie not in their respective genes, which are identical, but in their respective cytoplasmic computers. Differentiation and embryogenesis are processes managed by the cell's cytoplasm, not the nucleus. The DNA provides the reference library, but the cytoplasm does the heavy lifting (and heavy thinking) when it comes to assembling a body from a single egg.

Think back to the process of cloning. To clone another person like myself, I must take a nucleus from one of my skin cells, then take an egg from some willing female donor, remove the nucleus from that egg, and replace it with my own nucleus. Note that I have to implant my nucleus into an egg; simply implanting one of my skin cells into a uterus won't do the trick. The cloned egg and the skin cell have all the same genes, yet only the egg can initiate embryogenesis. Because they differ only in their cytoplasmic makeup, the difference between an egg and a skin cell must lie there, in the cytoplasm. The cytoplasm of the skin cell is perfectly suited for the tasks of a skin cell, but not up to the task of recreating the whole body.

To resort to the computer analogy once more, the nucleus contains a mass of information loaded onto DNA base sequences like so much data on a hard drive—the nucleus is our software library. However, some computational machine must access and implement this library. A hard drive is no good without a computer to read and act upon the information it contains. In the cell, the cytoplasm plays this role, accessing nuclear information selectively and for a narrowly defined purpose. Curiously, according to the calculations made by Manfred Spitzer in his book *The Mind Within the Net*, the informational capacity of the human genome is about 750 megabytes, roughly the amount that could be stored on a single compact disc. (If we take into account the fact that the modern CD was designed specifically to hold all of Beethoven's Ninth Symphony on a single disc, we see that the human genome holds as much data as the world's greatest symphony—armchair theologians can muse on this to their hearts' desire.)

The hard drive on my computer contains many programs, but right now I'm using only the word processor. Tomorrow, I may be using a game program, or maybe an Internet program. The hard drive may be the same, but the behavior of my computer will be different depending on what application I have open at the time. The hard drive is the computer's genotype, its germline, defining all that it can be. The software application is the phenotype, defining what functional characteristics my computer manifests at this moment. Presently, my computer is in the "word processor phenotype"; tomorrow, it may in the "Klondike solitaire phenotype." Same machine, same genotype, very different phenotypes.

Each somatic cell contains the same nucleus, the same genotypic hard drive; cells access and implement the nuclear software in different ways at different times. The cytoplasmic computer of a cardiac cell reads only that part of the genome dedicated to the cardiac phenotype; a skin cell accesses only that part of the genome dealing with skin cells. The nuclei may be portable, but the cytoplasmic computers are not. That's why, in order to clone myself, I must insert my hard drive into the cytoplasmic computer of an egg.

Here we arrive at the inescapable conclusion: that a gelatinous droplet of enzymes known as the cytoplasm harbors a gargantuan amount of intelligent machinery. This idea isn't new; biologists have long known that unique cellular behavior derives largely from the cytoplasm. What biologists don't know is how the cytoplasm does what it does. Where and how does the cytoplasm store the know-how needed to maintain the various phenotypes in my body? It has to be stored in the form of a landscape, on some E surface derived from the cytoplasm's chemical network. From a cybernetic point of view, there is simply no place else it could be. This idea isn't new, either. I suggested it years ago, as did others. Dennis Bray, writing for the journal *Nature* in 1995, also suggested that the cytoplasm might function like a chemical version of a "neural" network. Much as we have done here, Bray outlined the nonlinear, "switchlike" behavior of enzymatic reactions and described how these reactions could be exploited to assemble an intelligent device operating at the molecular scale.

Skeptics will point out that the cytoplasm isn't really "thinking" or doing anything else requiring true intelligence at all; it's merely popping in and out of predefined, albeit very complex, phenotypic states. The cytoplasm, they will argue, only follows a narrow set of instinctive behaviors, like an ant, becoming muscle cells and nerve cells in response to hormonal cues secreted during differentiation. Once the final phenotype is established, the cell slavishly and unthinkingly fulfills the role it has been assigned, little more than a serf. All of this is very true, but these arguments don't invalidate the intelligence potential of single cells. Serfs had minds; they just didn't use them very often.

Even serfs occasionally rebelled and showed a glimpse of their true intellects. Unfettered by the constraints of the predetermined landscape, cellular serfs can also become willful and cunning. Like dissidents escaping an oppressive regime, their formerly dronelike minds reawaken. In the body, these rebels carry a fearsome name: malignancy.

Cancer cells are often portrayed as good cells gone bad, the cellular equivalent of juvenile delinquents in a bad B movie. In the biological realm, however, there is neither good nor evil, only creatures striving to achieve their goals as best they can. Cancer reminds us that our bodies are social colonies, mammoth and uneasy alliances of many billions of self-minded cells. Such cells cooperate to form multicellular colonies solely because the body's vast symbiosis serves their own needs first, the needs of the colony second. The colony, for its survival, must keep cells in line by creating an environment that makes them happy in their various differentiated states; the body dupes its members into thinking that life as a muscle or a nerve cell is really their optimal path for personal survival.

But cancer cells defy the colony's edicts and revert to an undifferentiated state, similar in appearance (and behavior) to the pluripotential cells of the early embryonic period. During embryogenesis, pluripotential cells divide rapidly so that the embryo may grow. They also migrate to form organs, limbs, and eye buds, and they spur the development of new vascular channels to supply the enlarging creature with nutrients and oxygen. Likewise, cancer cells divide rapidly, migrate (metastasize), and spur the development of new blood vessels.

In the body's orderly landscape of phenotypic basins, cancer cells fail to land in any differentiated valley. They roll about like homeless nomads, reproducing aimlessly and producing limitless numbers of nomadic descendants. Left unchecked, they will choke the life out of their hosts.

How does this happen?

Cancer is a landscape failure, an error in differentiation. Somehow, a cell either pops out of its normal phenotypic attractor basin or never makes it there in the first place. The blame may rest with the cell, with the landscape, or both. An internal malfunction may prevent the cell from traversing the phenotypic landscape, like a coaster car slowed by rusted, useless wheels. Alternatively, the landscape may be distorted, a roller coaster with a buckled track that keeps cars from rolling downhill.

Cells communicate with their immediate neighbors (and with the body at large) via chemical signals. These signals range in size from giant protein hormones, like insulin, to simple molecules, like nitric oxide. Hundreds, perhaps thousands, of different signals continuously bombard every cell. These signals create a "hormonal melody," a tune composed of a myriad of chemical signals of various sizes, shapes, and concentrations. To stay differentiated, each cell must "hear" and recognize it own familiar melody. The phenotypic landscape becomes analogous to my memory for 1960s tunes, with different attractor basins for different melodies.

Recall that the brain, after hearing a few notes, can recognize an entire song. My brain's landscape was molded by experience; I wasn't born with the melody to "Proud Mary" imprinted in my cortex. The cell's memory for hormonal melodies is inborn, not learned, but it functions the same way. If a cell hears a certain hormonal melody, it instantly recognizes it and takes the melody as a signal to remain in its unique differentiated state. Liver cells stay liver cells because they recognize the liver melody playing about them. Scientists have been able to recreate rudimentary versions of a few of these tunes inside a test tube, convincing cells to switch phenotypic states outside of the body. The vast symphony of chemical signals enforces the body's phenotypic landscape and guarantees that cells will go through life with the same monotonous melody playing in their cytoplasmic noggins.

Here's the beautiful part of this arrangement: it's robust. In informational lingo, that means the body's system of differentiation is resistant to error and fairly immune to damage. Remember, machines that use landscapes for memory and problem solving are also good at pattern completion. They can recall a stored pattern even if it is altered, damaged, or otherwise incomplete. As in horseshoes, close is good enough for landscape-based thinking machines.

Those of us who use computers every day know that the type of reasoning used by microprocessors is not robust at all. If my computer makes the slightest misstep, it will malfunction or even crash entirely. Such a system would be a disaster for multicellular organisms, where mutations, viral infections, and other injuries are a daily if not hourly occurrence. Cancer occurs when the body plays the wrong melody or, much more commonly, the cell suddenly becomes tone deaf. Both events require major damage to the status quo. A musician would have to mangle "Proud Mary" (or I'd have to have one gigantic stroke) for my brain to misinterpret it—that's how deeply the melody has been carved into my brain's E surface. Similarly, for a lung cell to mistake the "lung melody" always roaring in its ears, it must endure an awful cognitive insult. Unfortunately for Tom, decades of tars and nicotine did just that; smoking transforms lung cells into the cellular equivalent of Alzheimer's patients. They forget the lung tune and began dancing to their own orchestra, a dance of death for Tom.

Mainstream biologists blame genetic alterations—mutations and/or viral manipulations of cellular DNA—for the cell's failure to heed the hormonal melody surrounding it. The most popular model is the "multi-hit" hypothesis, wherein three or more harmful mutations, all directed at the cell's differentiation machinery, are required to cause a cell to turn cancerous. This goes along with the landscape paradigm of differentiation; three or more harmful mutations in critical metabolic pathways are indeed a "major stroke" as far as a cell is concerned. A less robust system might be disrupted by only a single mutation.

Space doesn't permit a discussion of the myriad types of mutations and other DNA alterations that have been identified in cancer cells. In general, they involve defects in the cell's membrane receptors (the cell's "ears," essential for hearing the hormonal melody). The mutant receptors may be absent or malformed, rendering the cell deaf to certain hormonal "notes." Some mutations cause a cell to secrete excess amounts of growth hormones, saturating its own receptors and numbing the cell to the external melody; the cell yells at itself so loudly that it can't hear anything coming from the outside. Still other mutations affect the chemical pathways that connect membrane receptors to the cytoplasmic network inside the cell. To push the ear analogy still further, these mutations act like nerve deafness, garbling the signals the cell receives from the outside.

The genes that encode for the cell's ears, the receptors and receptor pathways needed for hearing the hormonal melody of differentiation, are called oncogenes (from *onkos,* Greek for "mass" or "tumor"). The word *oncogene* is misleading: these genes don't exist to cause cancer; they are normal genes found in every cell. However, functional damage to three or more oncogenes will indeed deafen the cell to the body's melodies and trigger cancer, hence the name.

Curiously, oncogenes were first identified by their ubiquitous presence in a certain class of viruses known as retroviruses. Retroviruses contain RNA, not DNA, and typically carry at least one gene swiped from a large multicellular species like us. The most famous retrovirus is the human immunodeficiency virus that causes human AIDS. Retroviruses use their purloined gene to alter the behavior of infected cells to suit their needs. Cancer-causing retroviruses carry mutant versions of human oncogenes because they prefer to reproduce in slightly deranged cells. After inserting their altered genes into a host's DNA, the viruses then set up shop in the damaged cells. If the infected cells go on to become fully cancerous, and we happen to die, that's of no concern to a virus, just as a zebra's death causes no sleepless nights for a lioness.

Not all cancers come from viruses; in fact, very few human cancers likely have a viral origin. Nevertheless, cancer-causing retroviruses pointed the way to the normal genes responsible for maintaining the differentiated state. Tom's cancer came from years of exposure to smoke-borne mutagens that eventually damaged multiple oncogenes in a few cells. These deafened cells, and their equally deafened progeny, now filled one lobe of his lung.

Perhaps a specific example of an oncogene will make all of this clearer:

Tom's cancer was a squamous cell carcinoma, a tumor arising out of the epithelial cells lining his airways. Epithelial cells are the body's boundary cells, forming the barrier between us and the outside world. These cells take the brunt of chemical carcinogens found in inhaled smoke. (*Karkinos* is Greek for "crab"; ancient pathologists likened cancer to a crab because of the way it crawls through normal tissues.) Epithelial carcinomas, which include cancers of the lung and breast, are the most common form of human malignancy, probably because of their exposed nature. One of the "notes" in the epithelial cell melody is supplied by a protein called epidermal growth factor, or EGF. EGF is one of the hormones that order a lung cell to stay a lung cell forever.

EGF binds to a membrane receptor on the surface of epithelial cells. The EGF receptor is really a large enzyme that straddles the cell's external membrane, with half of the enzyme inside the cell and half out. The enzyme's catalytic site faces inward, toward the cytoplasm, where it can interact with proteins inside the cell; the regulatory site, the site that turns the enzyme on and off, resides outside the cell. It is the regulatory site that binds the hormone EGF. When EGF binds to the receptor, the catalytic site is turned on and the enzyme helps the cytoplasmic computer to maintain the epithelial personality of the cell. Hormone receptors act just like human sense organs, translating external signals into a language comprehensible to the internal cytoplasmic brain. Our thoughts are, to a great degree, directed by our sensory input. The cell is no different in this regard.

The cell's EGF "ear" can be damaged or tricked in several ways. For example, the epithelial cell itself may secrete large amounts of EGF, activating its own EGF receptors. This is the case of a cell yelling its own tune so loudly that it can't hear any notes from the outside. A cell does not normally produce hormones that target itself, but cancerous cells frequently engage in this molecular version of masturbation. In a symbolic sense, masturbation really does make them go blind. (Or deaf—I may be mixing my metaphors here.)

The EGF receptor can also malfunction. There's a retroviral version of the EGF receptor, called v-erbB, which triggers a form of red cell leukemia, or erythroblastosis, in birds. This mutant receptor has a normal catalytic site but lacks the EGF-binding site. The receptor is stuck in the "on" position and is unresponsive to EGF altogether. Cells equipped with v-erbB have EGF tinnitus—they hear a ringing in their ears all the time.

As I've noted, one derangement alone is not sufficient to turn a cell cancerous. Cells are smart enough to ignore one false note in the melody. EGF is only one of the hormones involved in keeping lung cells sane; there are dozens of others. In vitro experiments suggest that at least three hormonal signals must be affected severely for a cell to drift dangerously away from its differentiated phenotype.

A key point here: mutations cause cancer because they alter the behavior of the cell's enzyme computer by depriving it of sensory input. As a network device, the enzyme computer is far more resistant to such errors than a personal computer. The cytoplasm, like the brain, is good at extracting useful information out of noisy data; a PC is not. Nevertheless, the enzyme computer can do only so much. In the face of overwhelming noise, it too will fail.

Surgeons cut out his diseased lung, but Tom's cancer popped up in his brain and abdomen. Cancer cells have their own script, and they follow it remorselessly. We treated the metastatic lesions with chemotherapy and radiation therapy, but the cancer cells fought back, activating enzymes to pump out the toxic drugs and to repair the genetic damage inflicted by our megavoltage X rays. Just as antibiotics reawaken the dormant intelligence of bacteria by accelerating the stresses on them, chemotherapy and radiation bring out the infinite resourcefulness of cancer cells.

In truth, cancer cells are more ingenious than bacteria. Bacteria have the greater power of numbers and a good deal more experience; they've been on this earth much longer than malignancies. Nevertheless, we have been able to go a few rounds with bacteria. Cancer, on the other hand, has us on the mat and keeps pummeling us. Most of the advances in dealing with solid cancers like Tom's have come in the form of earlier detection and treatment. Once they've spread, we have no more to offer their victims than we had in the Middle Ages. Squamous cell cancers like Tom's are particularly cunning; they seem to be chuckling at us as we launch our puny weapons against them.

On a hot summer morning, Tom's body reached complete equilibrium in his bed. His E surface flattened into one infinite tabletop and he headed for another meadow.

In another, more peaceful space.

So, how far have we come in this chapter? Far indeed. We have seen that most chemical reactions have fairly simple steady-state equilibrium behavior, too simple to act as components of any computational or memory storage device, and that computational devices must be built out of devices with at least two, if not more, steady states. We saw how chemical reactions can acquire two or more steady states if they are catalyzed by enzymes, particularly if the enzymes are subject to feedback effects, like the amplifiers in a computer flip-flop circuit.

From there, we progressed to networks of interconnected I/O devices, devices endowed with enormously complex equilibrium landscapes that can be used as problem-solving and/or memory devices. Moreover, we explored how the landscape approach to intelligence comes equipped with pattern recognition and robustness, both desirable properties in living systems.

Finally, we saw how an enzyme network, the cytoplasm, can use its landscape to store the various phenotypes needed to create even larger networks known as multicellular organisms, and how the landscape can fail if a cell suffers so much genetic damage that the enzyme network no longer recognizes the external hormonal melody. Jackson's idea of "switching" cancer cells back to normal cells might not be so far-fetched after all. The real solution to cancer will not be found in stronger poisons and deadlier radiation beams but in greater knowledge—specifically, greater knowledge concerning the cell's enzyme computer and how it interfaces with the rest of the body. To conquer cancer, we must understand the cybernetics of complex enzyme networks, a topic that has recently sparked significant research interest. Interestingly, cancer may give us our best window for seeing into the machinery of the cytoplasmic network. Cancer, a major cause of death, may enable us to understand what life is. What ultimate irony!

We have begun exploiting the body's differentiation melodies from a practical standpoint. For example, if a patient's white cell count falls too low as a result of chemotherapy, we can administer a hormone known as granulocyte stimulating factor (GSF), forcing more white cells to grow and differentiate in the bone marrow. By infusing GSF, we are, in effect, turning up the volume on the body's white cell melody. Of course, toying with these melodies for our own, unnatural purposes is risky. By administering hormones like GSF, we are acting like giant retroviruses, corrupting the melody and putting patients at risk for other cancers. In dying patients, however, this is a necessary risk.

The ultimate goal of oncology is to discover the miracle Jackson described, namely, some agent or combination of agents that will switch the cancerous phenotype back to normal. The best cancer therapy will not be one that kills the tumor, but one that merely shoves it back into a normal differentiation basin where it can harmlessly live out its existence. Research along these lines continues, with no clear breakthrough in sight. Unfortunately, we may be underestimating the problem. Recreating a lung cell's differentiation melody may mean rewriting a veritable Beethoven's Ninth in terms of hormonal notes and scales. We may not have the technology to do that for many decades.

And at this juncture, we must leave Tom, cancer biology, and the entire world of the living and safari deeper into the realm of the abstract to explore the behavior of networks in more detail. The trip will be quite unsettling, but eminently necessary. To know life, we must know more about social computation—the odd and wonderful phenomenon of network intelligence.

PART II
THE ABSTRACT

4

THE YANG
AND THE UM

*The organic and inorganic worlds are
both of similar nature and subject to the
same natural laws.*

—Leonardo da Vinci, *Notebooks,* circa 1518

For reasons even I don't fully understand, I decided to take up the Korean martial art of Tang Soo Do at the rather advanced age of forty. Like all traditional martial arts, Tang Soo Do stresses philosophy as well as fighting. (For people my age, Tang Soo Do stresses the hamstring muscles more than anything else, but that's another story.)

I learned that a fundamental principle of Korean philosophy—indeed, of all Oriental philosophies—is dualism, the belief that the world exists as a dynamic balance of polar opposites: good and evil, health and sickness, Republican and Democrat, and so on. To see the great importance of dualism to Koreans, we need only look at their flag. At the flag's center lies a circle, the Tae Geuk, emblematic of the world's split personality. An S-shaped line splits the Tae Geuk horizontally into two parts: the upper half (the Yang) is red, symbolic of heat and passion, and the lower half (the Um) is blue, signifying coldness and calm. Surrounding the Tae Geuk are two pairs of opposing line drawings representing sky and earth, fire and water.

The boundary between Yang and Um is wavy, not straight, and it has been fashioned this way to symbolize the ever-fluctuating interface between competing forces. The world is not statically partitioned into hot and cold, dark and light, weak and strong, but rather, the boundaries between opposites exist in a state of constant flux and, in some cases, may not be easily defined at all. The wavy line signifies the uncertainty and noise that permeate our world. As will be seen, the biological world is also dualistic, filled with many opposites (and also filled with noise); but before we can return to the realm of the living, we must travel to the world of the abstract. This chapter and the next two will lay the conceptual groundwork for understanding the intelligence of living systems. I will try to make the ensuing material as intellectually digestible as I can. As Einstein once observed, science should be made as simple as possible, *but no simpler.*

In the spirit of the Yang and the Um, let's consider a dualism that pervades nearly every aspect of modern science: the dualism between the world of the small and the world of the large. Often, these two worlds look irreconcilably different. For example, quantum mechanics dominates the behavior of atoms and gravity rules the behavior of planets. Although both describe the physical world, quantum physics and gravitational physics have little else in common. Likewise, air is simply a large collection of molecules bouncing about like billiard balls, yet a group of 10^{24} discrete molecules, when viewed en masse, behaves like a gas, and gases don't look anything like collections of billiard balls. This leap from the small world of atoms and molecules to the large world of gases and planets, like the alchemist's leap from lead to gold, appears mysterious and without any intuitive foundation.

Here again we encounter the now familiar concept of emergent properties. An emergent property appears only on the *large* scale, a consequence of the bulk action of many individuals interacting on a *small* scale. Emergent properties embody the proverb that the whole is often more than the sum of the parts. Gravity is an emergent property of atoms aggregated into planet-sized lumps, yet we see no hint of gravity when studying isolated atoms. Gases are emergent properties of large collections of atoms or molecules, even though atoms don't resemble fluids or gases in the least. In both cases, a Lilliputian observer, armed only with an understanding of the atomic world, would be astonished to learn about emergent things like gravity and air. To understand life, we must understand *emergence*. In particular, we must understand the peculiar forms of emergence manifested by collections of organic things.

The behaviors that emerge from any large group are determined by two factors: (1) the nature of the objects making up the group, and (2) the way objects within a group interact with one another and with the outside world. Obviously, a group of nerve cells manifests one set of emergent properties (the properties of a brain), and a group of room-temperature oxygen atoms another set (the properties of a gas). Also, a group of water molecules at 0 degrees C will have the emergent properties of a solid, while a group of water molecules at 30 degrees C will have the emergent properties of a liquid.

In the case of nerve cells and oxygen atoms, the emergent properties of the two groups differ because the *character* of their members differs. In the second case, the members (water molecules) are identical in both groups, yet the groups exhibit different emergent properties because the *interactions* among members vary according to temperature. In ice, the interactions among water molecules are strong; in liquid water, the molecules interact more weakly.

It's relatively easy to understand how two groups comprised of different members will behave differently—obviously, a group of nerve cells won't act like a group of atoms. On the other hand, it's a little more difficult to understand how two groups comprised of identical members may manifest vastly different properties depending upon how those members interact. Nevertheless, this point can't be stressed too strongly: *the nature of the interactions among group members becomes key to understanding the group as an emergent whole.* Although the neurons making up a chimp brain are identical to human neurons, chimps aren't humans. The pattern of the brain's neuronal connections distinguishes a Harvard professor from the inhabitants of a zoo. We now have the technology to grow human brain cells in tissue culture, making it feasible to toss several billion human nerve cells into a beaker and keep them alive and functional. But a beaker of nerve cells doesn't make a brain; the proper connectivity just isn't there.

Consider another, somewhat grislier thought experiment: take my body and separate it into individual cells (theoretically, this can be done without killing any of the cells). Now, place each cell into a separate test tube filled with life-sustaining nutrients. Am I still alive? Technically, no part of me has died, not even a single cell, yet I have ceased to exist. In effect, I've died, yet all my cells survive. That entity known as my life is defined not by the cells themselves, but by how the cells are connected together. Likewise, at the cellular level, the thing called life is not a matter of enzymes and nucleic acids, but a matter of how those molecules interact to form an emergent group.

How things connect and how they interact with one another determines what sort of emergent systems they will become on the large scale. The million-dollar question becomes: what sort of connectivity turns enzymes into life, lymphocytes into an immune system, and nerve cells into brains? And are the patterns of connectivity among enzymes, lymphocytes, and nerve cells the same? These are the questions we must address in this chapter and the next two. Both life and intelligence (which aren't separable entities, in my opinion) are emergent properties of large groups comprised of nonlinear "things." What those things are and how they interact determine the emergent character of the group. Enzymes make cells; cells make organisms; organisms make ecosystems. To understand life itself, we must understand how organic things are assembled into intelligent aggregates.

Intelligent aggregates, whether comprised of transistors, molecules, cells, or whole organisms, must be connected in a way that makes them capable of learning and analyzing patterns of complex information. In the last chapter, I referred to such aggregates as networks. The term *network* comes from computer science and, at first glance, it seems almost too trivial a word, given the profound concepts involved here. As will be seen, networks aren't some software fad; rather, they represent an entirely unique paradigm of computation—a fundamental way of manipulating information. We could apply a more poetic phrase to network devices (landscape thinkers?), but we're stuck with the more mundane computer science lingo, whether we like it or not. Computer scientists were among the first to explore the basic mechanisms of intelligence in any depth, and so they get the naming rights.

To understand how networks do what they do (and why they're so fundamental), we'll explore one type of network model in detail: the Hopfield network. Before examining this network, though, we'll need a little detour into computer history. While this book isn't about computers, we can't understand the nature of intelligence without delving just a bit into the nature of computing.

◎

John Hopfield, a physicist, is one of the leading architects of modern *connectionism,* the study of interconnected networks. Connectionism, also called network theory, is a branch of the broader discipline known as Artificial Intelligence, or AI. The name *AI* was rather poorly chosen, because it implies that the world contains two distinct species of intelligence: natural (the intelligence of life) and artificial (the intelligence of machines). In computer parlance, "natural" intelligence comes from wetware (living cells) and artificial intelligence from dryware (electrical circuitry). Ironically, one of the underlying tenets of AI research is the universality of computational processes. AI acknowledges that both wetware and dryware use similar, if not identical, methods to achieve intelligent status. Thus, AI rejects its own name; there can be no theoretical distinction between artificial and natural intelligence. Intelligence may be *implemented* in either living organisms or human-made electrical devices, but it's still the *same* entity as defined by a common set of computational and logical rules.

Our old friend Alan Turing (of the Turing test) hypothesized that all intelligent machines are simply variations of a universal computing machine made from "on-off" switches lined up in a row. His hypothetical computer was dubbed the Turing machine, and it exists only in the abstract. Oriental philosophers would be pleased to know that intelligent machines can be assembled using only two opposing states, 1 and 0, yes or no, on or off—the ultimate expression of Yang and Um.

Turing's machine presumes that intelligence can be described in general terms without reference to a particular physical device. The machine can be constructed using any bi-stable device, alive or dead, electrical or chemical, large or small. For the machine to be intelligent in the Turing machine sense, only its general *architecture* matters; its precise construction becomes a secondary issue, mere engineering details. As it turns out, living systems aren't

Turing machines, at least not as Turing envisioned them, but his basic idea still holds even for biological devices: intelligent systems can be built in many ways—immune systems, bacterial colonies, brains—so long as they share common architectural features. Unfortunately, models like the Turing machine are just theoretical constructs. In the real world, more than abstract theories are needed to fashion intelligent devices.

Early computer scientists seeking to build an intelligent machine looked to the human brain for additional architectural clues. Anatomists have long known that the brain consists of roughly ten billion neurons densely wired together and, although the human brain contains several neuronal subtypes, the vast majority of neurons look and act pretty much the same. Consequently, researchers sought to build a cognitive machine by interconnecting a large number of identical components in different ways. But how could they simulate a living nerve cell?

In 1943, neurophysiologist Warren McCulloch and mathematician Walter Pitts noted that real neurons behave much like on-off switches—everything boils down to those pesky little bistable devices. They proposed a simple model of the nerve cell that can be implemented using electrical components. Their model, called simply the McCulloch-Pitts neuron, is still used today. The McCulloch-Pitts neuron receives multiple inputs, from the environment and/or from other neurons, then adds those inputs to obtain a single sum. If that sum exceeds a certain threshold value, the synthetic neuron switches "on" and emits a constant output. If, on the other hand, the total input falls below the threshold value, the neuron stays "off" and emits no output. The threshold value determines what total input will flip the neuron's switch. This behavior does resemble that of real nerve cells to a certain extent, as will be seen.

The no-frills version of the McCulloch-Pitts neuron has only two tasks: (1) add up the inputs to achieve a total, and (2) decide whether that total crosses the threshold for producing output. The neuron can be souped up—for example, by giving it more sophisticated input/output behavior—but the simple "on-off" version will do for now. If this sounds too much like pure computer science, think again. Our immune system's lymphocytes (see chapter 2) act much like McCulloch-Pitts neurons, adding up their antigen input and deciding whether that input crosses the threshold needed to trigger cellular reproduction. And, as was seen in chapter 3, enzyme-catalyzed reactions also resemble McCulloch-Pitts neurons. In fact, all living systems, from cells to immune systems to brains, contain some version of the simple McCulloch-Pitts device.

The first connectionists began toying with network machines built from many McCulloch-Pitts neurons and seeing what these ersatz brains could accomplish. In truth, it was not much, at least not at the time. Flaws in their methodology hampered the first connectionists and, although they met with some early successes, they couldn't keep up with the evolving field of digital computing. Unlike network devices, digital computers aren't designed to mimic the brain explicitly. Instead, they use a sequential set of arithmetic and logical rules, called a *program,* to achieve practical results. Digital computers are "ruled-based," in that they must scrupulously follow a set of rules laid out by their human programmers.

Here we encounter one of the chief differences between the divergent fields of network and digital computing: *networks invent their own rules as they go, while digital computers need to have the rules spelled out for them by programmers.* (How networks make up their own rules will be explained in due time.)

The Turing machine is a rule-based device—in fact, it's the ultimate rule-based device, a godlike thing programmed with every conceivable rule. The imperfect physical realization of a universal Turing machine—the modern digital computer—is also called a von Neumann machine, after the fabled mathematician and computer pioneer John von Neumann. Early in its short history, AI researchers split into two warring factions: those, like von Neumann, who favored rule-based (digital) machines and those who favored connectionist (network) machines. As it turns out, this schism wasn't of human making. The two camps—rule-based computers versus networks—represent two separate paradigms of intelligence, just as quantum mechanics and gravity represent two separate paradigms of physics. Neither paradigm is "right" and neither is "wrong"; each has its own unique advantages and disadvantages. (Some experts contend that rule-based and network-based computations are, in a deep sense, merely different "versions" of the same thing, just as gravitation and quantum physics are different manifestations of a mythical "unified field theory." However, this theoretical controversy needn't concern us here.)

Our present technology favors rule-based machines, and nearly every computer we interact with today is of this type. Life, on the other hand, prefers the connectionist approach, and for an obvious reason: rule-based machines need a programmer, networks do not. Theology aside, no one programmed my brain to play tennis—it learned the skill all by itself, through experience. The microbial mind used no sequential set of instructions to decipher its antibiotic threats. Networks can solve problems without external programming, but the rule-based computers of von Neumann and Turing (example: my laptop PC) can't. How do networks learn all by themselves? Read on, for therein lies the meaning of life itself.

In the early years of the AI wars, things looked bleak for the connectionist camp. Faced with stiff competition from rule-based computers and producing only lackluster results of their own, the connectionists struggled to find a niche in AI. In 1969, network theory was apparently killed outright by the landmark book *Perceptrons,* by Marvin Minksy and Seymour Papert, two of AI's leading pioneers. Minsky and Papert demonstrated that one of the most popular models built from synthetic neurons, the perceptron, was theoretically incapable of modeling complex intelligent behavior.

The perceptron was created by Cornell University psychologist Frank Rosenblatt in 1959. In its original incarnation, the perceptron consisted of a bank of photocells connected to a bank of synthetic neurons. The photocells acted as an electronic retina, and the neurons simulated the brain. The perceptron was a perception device, hence the name, and Rosenblatt showed that it could learn to recognize the entire alphabet. More important, the perceptron could accomplish this without any preconceived set of rules, unlike a digital computer, which can't do anything without detailed rules. Connectionists particularly liked the perceptron's ability to learn on its own, since that's how living things learn, too.

Unfortunately, the earliest perceptrons couldn't do much more than learn a handful of patterns, as Minsky and Papert proved. Consequently, early hopes that network machines could outdo rule-based digital computers soon faded. True, digital computers need programmers, but humans could perform that task for them. Whether living networks have a "god" can be debated indefinitely, but digital computers do have a god, and that god is us. The practical successes of digital computing soon outpaced connectionism. The 1970s saw a decline in connectionist fervor, and funding for network research soon dried up.

As network theory waned, interest in rule-based computers exploded, thanks to the advent of transistors and integrated circuit technology. Prior to the transistor, von Neumann's rule-based machines suffered from a crippling lack of speed and reliability.

Rule-based computers are also called *serial processors,* because they perform only one operation at a time. (An "operation" is an arithmetic or logical step in a program, such as $2 + 5 = 7$, or "Is $x > y$?") All operations are carried out, or "executed," by a single central processing unit, the CPU, which grinds through programs consisting of long series of operations to be carried out sequentially, in a prescribed order. No matter how big or sophisticated the serial computer may be, its CPU still can do only one thing at a time. Moreover, in serial machines, the CPU and memory storage areas occupy different parts of the same machine; as such, the speed of the serial machine is severely limited by the rate at which data flow between the memory and the CPU. These "computational bottlenecks" make all serial computers intrinsically slow, regardless of their particular construction. Their sluggishness can be overcome only by increasing the speed at which the CPU executes individual operations—if you can do only one thing at a time, then do it very fast. (Even the casual computer shopper now knows the importance of processor speed.) But significant advances in CPU speed weren't possible until the invention of the transistor.

It should be obvious by now that all intelligent machines— whether alive or dead, rule-based or network-based—rely upon some form of switch; a machine's speed is then determined by how fast its switches can go from on to off and vice versa. Think of Morse code, another binary system for encrypting data: the speed of data transmission is determined by how fast the telegrapher can push the key up and down, on and off. The faster the switches move, the faster the machine works. The first digital computers, the electromechanical machines of the 1940s, used mechanical relays swiped from heavy industry. Relays are notoriously slow devices, taking up to a second to switch polarities, and the monstrous computers made from them looked more like steel mills than thinking machines as they performed their clattering operations at a snail's pace.

Later, vacuum tubes replaced relays. Electronic tubes are much faster than mechanical relay switches—tubes can flip on and off thousands of times each second—but they are still painfully slow by modern standards. To the early computer engineers, tubes posed an even bigger problem: they were unreliable. Tubes contain a filament that heats and cools as the device switches on and off, and, as anyone who lived through the tube era of televisions and radios knows, tubes burn out with alarming frequency (and usually at the worst possible time). The legendary ENIAC, constructed at the University of Pennsylvania under the guidance of von Neumann himself, contained almost 18,000 vacuum tubes. If only one tube malfunctioned at a critical time, the machine's entire output would be ruined.

Fighter pilots have a motto: speed is life. Like pilots, computer scientists, especially those dealing with serial machines, highly value speed. Unfortunately, tubes and relays posed insurmountable physical barriers to further advances in computational velocity. Even worse, tube technology, which couldn't be easily miniaturized, placed an upper limit on a CPU's complexity. A von Neumann computer made of 100,000 tubes? Perhaps—if you have an airplane hangar to spare. But a million tubes? Impossible.

In the late 1950s and early 1960s, attention turned to networks as a possible solution to the serial computer's technical limitations, because connectionist machines have the potential to be the fastest computers of all. Networks aren't serial processors, but *parallel processors*. While serial processors have only a single complex CPU, networks contain many simpler "neurons" making small decisions simultaneously ("in parallel"). A computer with a thousand neurons does a thousand things at a time, not just one. Parallel machines thus avoid the terrible one-operation-at-a-time bottleneck that so cripples serial machines. And, because memories are saved in the network at large, as opposed to the separate memory banks of serial machines, parallel machines aren't further crippled by having to transfer data back and forth from a distant storage area. Parallel computation thereby eliminates the need for fast switches— a good thing, considering that the human neuron is a very slow switch (much slower than the average vacuum tube). The brain is a marvelously fast machine because of its parallel design.

Their inherent speed advantage made connectionist machines look very appealing in the 1950s and 1960s, the era of very slow switches. Networks are also more reliable than serial machines. In a serial machine, one component failing within the CPU spells certain doom. Networks, in contrast, are structurally robust. In a large network, the occasional neuronal malfunction won't hurt overall performance. (Consider that the work of a network is distributed among many simpler elements, making each element less important to the final outcome.)

Before artificial networks could gain any traction, however, advances in semiconductor technology cured the problem of slow, unreliable switches. Transistors can flip on and off in a billionth of a second because the switching is done at the molecular scale, not at the macroscopic scale of vacuum tubes and relays. Also, transistors, which have neither moving parts nor hot filaments, almost never fail—and they can be made vanishingly small. Suddenly, the computational bottleneck of serial digital computers no longer seemed an insurmountable barrier. Furthermore, enormously sophisticated CPUs could be designed without fear of component failure. ENIAC's 18,000 glowing, failure-prone tubes filled an entire room, but 100,000 or more immortal transistors can now be

packed onto a single chip the size of a postage stamp. Interest in serial machines began to rise, while interest in perceptrons and other
parallel machines began to wane even further. With the technical
limitations of serial computing solved, the industrial impetus to explore the uncharted realm of parallel computation vanished.

By the late 1960s, the theory of serial computation had advanced much further than the theory of parallel computation and
serial machines were becoming faster, cheaper, and more sophisticated all the time. Thus, connectionism as a practical discipline
was dying a slow death even before Minsky and Papert plunged a
dagger into its heart. The serial, rule-based way of computing
soon became the standard modus operandi of the modern computer, and it remains so to this day. In fact, serial computing became so dominant that many researchers began to believe that it
was the *only* pathway to true intelligence. This mistaken belief
hampered our understanding of biological intelligence, since biological intelligence bears no resemblance to the intelligence of
digital machines. The vast differences between how cells and microbial colonies "think" and how serial computers "think" further
contributed to our prejudice against biological systems, leading us
to underestimate their true intellects.

After the invention of integrated circuitry, networks of
simple-minded McCulloch-Pitts neurons quickly gave way to the
muscular microprocessor CPU and connectionism entered a
decade-long Dark Age. The field would lie dormant until Hopfield, a connectionism Leonardo, initiated a Renaissance by pub-
lishing a short treatise entitled "Neural networks and physical
systems with emergent collective computational abilities" in 1982.
Hopfield's work helped ignite fresh interest in network models of
intelligence. He discovered what biology had known for billions
of years: if you want to be *really* smart, and *really* fast, then parallel processing, not serial processing, is the way to go. Rule-based,

serial machines make nice countertop appliances, but only networks can be *alive.*

◎

Before proceeding, let's summarize:

Intelligent devices fall into one of two classes, depending upon their mode of "computation":

1. **Rule-based devices** (also called serial processors). These devices require extrinsic sets of rules, or programs, which the devices execute one at a time. Because they do single operations sequentially, serial processors are intrinsically slow (although rapid electronics can compensate for this flaw). Rule-based devices must also rely on a single CPU and are thus prone to catastrophic failure. On the plus side, serial machines are extraordinarily accurate and, when they are working properly, relatively free of noise and errors.

2. **Network devices** (also called neural networks, parallel processors, parallel-distributed processors, or connectionist devices). Networks distribute their computations over many simpler devices ("neurons") and can make up their own rules without the need for external programs. The parallel nature of their construction makes them very fast, even when their component processors are relatively sluggish. In addition, they are quite robust, in that they can tolerate some injury without catastrophic failure. Unlike serial machines, however, networks aren't particularly precise and can be plagued by noise.

Note the Yang/Um dichotomy of these two computational paradigms: serial versus parallel; slow versus fast; rule-based versus rule-independent; injury-prone versus robust; noise-free versus noisy. Note also that these two classes represent *abstract paradigms of computation,* not specific physical machines. My toaster must obey the laws of thermodynamics, but it doesn't define those laws, which exist only in the abstract. Likewise, my laptop operates in the rule-based paradigm of intelligence, but it doesn't define that paradigm. What we have are two different theoretical frameworks for describing intelligent behavior.

If we assume that life is an intelligent device (a safe assumption, I think), then it clearly must fall into the network category. Life must make up its own rules as it goes; it must be robust and, since no living "switch" can work as quickly as a silicon transistor, it must think quickly. Although networks don't work as precisely as serial machines, living systems don't need to find mathematical answers correct to ten decimal places. Moreover, as we'll see later, the noise inherent in networks can be harnessed for productive purposes. Thus, life is a network and, difficult as network theory may be, those of us who wish to understand life (and its inseparable companion, intelligence) *must* understand something about networks. From here on, we will begin to think and talk entirely in terms of network organization. If you review the previous chapters, it will be apparent that I've been talking about networks all along (networks of bacteria, lymphocytes, and enzymes), although I wasn't explicitly using network terminology.

The first artificial connectionist networks, including the perceptron, had a rather simple design that proved to be their downfall. The original perceptrons arranged their neurons in two neatly defined rows, one row for input and one for output, like ducks in an arcade game. As AI researchers would later learn, smart networks need more than two precise rows of neurons. The perceptron's neurons were like the British Redcoats during the Revolutionary War, who fought battles in the same rigid formations regardless of the local terrain. Smart networks behave more like the ragtag American rebels, who preferred irregular formations that could be fluidly adapted to the lay of the land. Like the Redcoats, perceptrons had an overly rigid and simplistic architecture that hampered their usefulness.

Early AI researchers gravitated to "Redcoat" versions of networks because orderly networks were easier to model and simpler to build. In orderly networks, neurons are neatly grouped and their interconnections follow a regular pattern. Although they may be easier to understand, however, orderly networks aren't very interesting. Disorderly networks, containing highly irregular patterns of connections, prove to be vastly more useful.

To understand complex, irregular networks, John Hopfield looked not to the brain, but to inorganic substances like magnetic materials and spin glasses (substances containing particles with different directions of spin). Hopfield argued that certain nonliving materials could perform rudimentary feats of collective memory by storing information in the form of stable states:

> *In physical systems made from a large number of simple elements, interactions among large numbers of elementary components yield collective phenomena such as stable magnetic orientations.... Do analogous collective phenomena in a system of simple interacting neurons have useful "computational" correlates? For example, are the stability of memories, the construction of categories of generalization, or time-sequential memory also emergent properties and collective in origin?*

> Hopfield, 1982

As Hopfield knew, inorganic materials will resemble simple "parallel computers" if they contain huge numbers of bi-stable entities interacting simultaneously. Magnetized materials, for example, contain tiny, bi-stable magnetic crystals that act as switches, flipping their north/south orientations in response to the ambient magnetic forces exerted by surrounding crystals. The crystals behave like tiny McCulloch-Pitts neurons, summing the magnetic forces around them (input) and determining their orientation (output) accordingly.

Large numbers of magnetic crystals or spinning particles, taken as a collective whole, possess a limited number of stable states. In other words, *certain physical substances such as magnetic materials and spin glasses manifest E surfaces with multiple hills and valleys.* If we "initialize" a collection of magnetic crystals near one of its stable valleys, the entire ensemble will roll into that valley and remain there until another force pushes it out. If we use this property to align collections of magnetic crystals on a piece of film, the crystals will stay collectively aligned until we magnetize them into another stable state. This is the basis of reversible data storage on a magnetized computer disc.

Now consider the spin glass, a mixture of spinning particles with both attractive and repulsive tendencies toward one another. The typical spin glass is a metallic alloy consisting of a nonmagnetic metal containing a small amount (usually less than 10 percent) of some magnetic material. For example, alloys of gold (nonmagnetic) and iron (magnetic) can show spin glass behavior. In a heated spin glass, the thermally agitated metal particles will be randomly aligned. As the spin glass cools, the particles seek to align their directions of spin to match their surrounding particles. However, the whole glass won't agree on a single direction of alignment. Instead, random islands of similarly aligned particles will emerge, creating a complex mosaic structure wherein each island represents a regional steady state for the particles trapped inside. The whole glass becomes an irregular tapestry of different steady states, an E surface landscape. Unlike most solid-state crystals (which contain periodic, or repeating, structures), spin glasses exhibit a complicated "aperiodic" structure. (Physicist Erwin Schrödinger once called life an "aperiodic crystal." He didn't realize it at the time, but he was comparing life to a spin glass, a rather prophetic statement as it turns out.)

Hopfield sought to exploit the tendency of nonlinear physical ensembles, like spin glasses, to organize spontaneously into complex patterns *by devising some way of storing meaningful information in those patterns.* Instead of letting a spin glass simply condense into random islands of aligned particles, Hopfield looked for a way of imprinting, or *mapping,* external data onto the mosaic as it cooled by assigning some information to each regional steady state arising inside the spin glass.

The last chapter showed that an E surface with more than one valley can serve as a form of memory device. Why is this so? Because memory devices must have at least two *relatively* stable steady states to store information, and most E surfaces fit the bill perfectly. Relative stability means that a state is stable enough to store information reliably, yet unstable enough for some outside force to switch it to another stable state in the future.

For example, rural mailboxes typically have a little mobile flag on their side. When we position the flag vertically, we tell our postal workers that we have placed outgoing mail in the box. When the flag rests horizontally, the box is empty. We use the flag as a memory device, a way to store information about the status of the mailbox. The flag must "remember" its position until the postal service arrives—it must be stable when it's either up or down. A loose flag, one prone to drifting from the upright position, is useless. If it can be blown down by the slightest gust of wind, it can't be trusted to store information. But the flag can't be infinitely stable in one position, either. A flag welded permanently into place is equally useless. We need the happy medium: a device stable enough to hold onto its memory in the face of random noise, yet *unstable* enough to permit easy transitions between alternate states. We want relatively stable states that will hold fast until *we* decide to change them. To build intelligent machines, we need a form of reversible memory, one that balances the Yang of stability with the Um of instability.

The undulating valleys on an E surface achieve this balance. Wagons roll to the bottoms of valleys, but we can get them out again with modest effort. This makes the E surface, which is a property of physical systems like spin glasses, ideal for storing memories. Hopfield realized this and sought a way of implementing the E surface concept in a computational machine by designing models of networks patterned after physical systems. But instead of spinning particles or magnetic crystals, Hopfield used electrical McCulloch-Pitts neurons. By connecting the neurons together in a certain way, he found that they would align into random pockets, much like those formed by particles in a spin glass. These pockets, in turn, would fashion a synthetic E surface with energy valleys that could be used as containers for stored memories. It would then be just a matter of finding a way of assigning specific memories to specific valleys in that E-surface landscape.

Instead of arranging neurons in tidy, symmetrically connected rows, Hopfield connected each neuron to every other neuron to make one massive web. At first glance, this form of network hardly seems a realistic representation of a physical system. In a large piece of magnetic material, one crystal can't interact with all other crystals simultaneously, at least not with equal strength. So Hopfield added another twist: *the "strength" of the interaction between any two neurons varied.* Again, this is analogous to a physical system: a magnetic crystal or spinning particle could be influenced by any other crystal or spinning particle, but not to the same degree. Crystals nearest to one another would interact strongly, while those separated by a great distance would interact weakly. (The weighting of connections wasn't Hopfield's idea, but his use of this concept proved particularly effective.)

Within a network, information between neurons flows in one direction: one neuron's output is sent, via a "connection," to another neuron, which receives it as input. One neuron may be connected to hundreds of others at the same time, but the information flow between any pair goes over a single connection. Think of the phone system: one phone can access any other phone in the world, but there is, technically, only one connection between any two phones. But why use a generic word like *connection*? Why not just call them wires?

Because a network isn't always an electrical device. In biological systems, electrical networks (brains) were the last type to evolve. The more abstract term "connection" is appropriate here; the general theory of networks, like the general theory of serial computing, can be defined in abstract terms without regard to any particular machine. The brain and the immune system are two imperfect realizations of the one "ideal" network, but there are many others. The cytoplasm is also a network, but the connections within a cell don't resemble electrical wires. Similarly, lymphocytes form networks using molecular signals in place of wires. We'll revisit this in much more detail in the next chapter.

The critical point here is this: not all connections in a network are equal. The relative strength of a connection within a network is called its *connection weight*. The weight is a number to be multiplied by a neuron's output before being relayed to another neuron as input. For the purposes of illustration, let's extract two neurons from a network, any network, and call them neurons A and B.

Assume that A's output is connected to B's input with a weight of w. We'll arbitrarily assign w a value of 0.5. Now, if A emits an output of 1.0 and sends it in B's direction, B will only receive a value of 0.5 ($1 \times 0.5 = 0.5$). The connection weight of 0.5 reduces A's output before it reaches B. If we lower w to 0.1, B receives an input of 0.1. As w approaches zero, A's ability to relay information to B gets progressively smaller; at $w = 0$, A can no longer transmit any input to B whatsoever, since zero times any number is zero. A large w means that two neurons are strongly connected, a small w means they are weakly connected, and a w of zero means that the two neurons can't communicate with each other at all.

The connection weight isn't a property of neurons but a property of their connections. The weight affects only the information transferred between two specific neurons. In the example above, the weight, w, affects the output of A only when it is relayed to B. The output of A may simultaneously be sent to many other neurons, and it may have stronger or weaker effects on those neurons depending on the strengths of their individual connections to A.

Again, think of the phone system. When I call my local neighbors, they have no trouble hearing me. When I call London, however, my voice will not be as loud. The problem lies neither with me nor the Londoner at the other end—we both speak and hear normally. The problem lies in the connection: a local connection is "stronger" than an overseas connection. By measuring the loudness of my voice at both ends, I could actually calculate the "weight" of my overseas phone connection. If a Londoner hears my voice at only 30 percent of my speaking volume, for example, the Pittsburgh–London phone connection has a weight of 0.3. At a weight of zero, the Londoner wouldn't hear me at all.

(For the more mathematically inclined, the complete set of all connection weights in a network is called the *weight matrix, w_{xy}*, where x and y are any two neurons in the network. The connection between A and B in our example above would be written w_{ab}. I'll use the phrase *weight matrix* when referring to the connections inside a network taken as a whole, but we needn't concern ourselves with the mathematics of matrices in any more detail, so breathe easy.)

Let's view the meaning of connection weights in more human terms:

A young college student is deciding whether or not to get married. No matter how hard he deliberates, he can't decide by himself, so he solicits the advice of his mother and father, his sister, five of his fraternity brothers, his priest, and, finally, a local bartender. He is now a McCulloch-Pitts neuron in a human network, possessing two stable states: marry and not marry.

The student is connected to ten other neurons (the ten people listed above), which feed him input in the form of a binary opinion: yes (get married) or no (stay single). We should take note here that the concept of neuron, like the concept of connection, is a generic one. A "neuron" here can be any decision-making, bi-stable entity, from a simple switch to a complex nerve cell—even a human being.

Our student solicits the ten opinions and adds them to determine a single input sum; he then decides whether that sum crosses his threshold for "switching states" from single to married. Let's arbitrarily define his triggering threshold as two opinions. In other words, if a total of two or more opinions vote yes for marriage, he will marry; otherwise, he'll stay single.

After polling his fellow neurons, he finds that four people say yes, he should indeed tie the knot. It sounds like an open-and-shut case. If he awards all opinions the same weight, then the student must marry, since four opinions clearly crosses his threshold of two yes votes. But he *doesn't* give all opinions equal weight. Who does? Not surprisingly, our student awards the greatest weight to the opinions of his mother, father, and sister, a lesser weight to the opinions of his fraternity brothers and priest, and no weight at all to the bartender (whom he met just five minutes ago). In network theory, he's "highly connected" to his family, "weakly connected" to his fraternity brothers and priest, and totally disconnected from the meaningless advice of his local barkeep.

In a mathematical network, actual numbers must be assigned to the connection weights between the student and his ten opinion-givers, and so let's make up some values:

Mother–student connection	=	1.0
Father–student connection	=	0.8
Sister–student connection	=	0.5
Priest–student connection	=	0.3
Frat brother–student connection	=	0.2
Bartender–student connection	=	0.0

Thus, if his mother votes yes, she counts as one full opinion. If a frat brother votes yes, however, he counts as only two tenths of an opinion. A yes vote by mother, father, and sister (total opinion input = 1.0 + 0.8 + 0.5 = 2.3 total opinions) would trigger him into the married state, but mother, priest, frat brother, and bartender (1.0 + 0.3 + 0.2 + 0.0 = 1.5 total opinions) would not. In the latter case, four out of ten people voted for marriage, yet he stayed single. The weights ultimately determine his fate.

We can use this silly example to gain a deeper understanding of connection weights. First, notice that the weight is a measure of the relationship between two people, not an intrinsic measure of either person taken separately. The mother–student weight is large because of their unique relationship, not because the mother is an expert on marriage. If she rendered the same opinion to a total stranger, the stranger would likely ignore it, making the mother–stranger weight zero. Again, weights are a property of the unique connection between two neurons, not a property of the neurons themselves.

Second, keep in mind that connection weights aren't fixed in time. At present, the student gives no weight to the bartender. Two years from now, the bartender may be his best friend and have a weight equal to that of his father or sister. Conversely, the student may become estranged from his mother over that same period of time, reducing her weight significantly. The relationships among neurons in a network, like those among people, are in constant flux.

One final point: the student has his own opinion about marriage. When other human neurons, like his frat friends, ask his opinion, he will tell them and they will weight that opinion accordingly. Every neuron receives and doles out opinions simultaneously. And opinions, like connection weights, vary over time. The student's mother may be for marriage today, against it tomorrow. The mother's opinion is a dynamic entity shaped by her own experience and influenced by other human neurons offering their opinions. We now glimpse how a complex network emerges: each neuron seeks opinions from others, weights them, adds them, makes decisions based on them, and then feeds their opinions back into the network.

If the concept of weights is still vague, go back and reread the last few pages. The soul of any network lies in its connection weights. The weights define how the network is connected and how its component "neurons" interact. Weights make the difference between a lump of random nerve cells and a brain, between a beaker of enzymes and a living amoeba. The behavior of the weights is the stuff of Victor Frankenstein, the spark that animates the dead machine. Weights are everywhere in biological systems: the interactions between proteins are weighted, as are the interactions among cells, plants, ants, even people. The role of weights must be very clear. From here on, *weights mean everything.*

Let's now return to a real version of a Hopfield network, one made from electrical McCulloch-Pitts neurons. Every neuron in the Hopfield network must constantly decide whether it should stay in its present state—on or off—or switch to the opposing state. To make this decision, the neuron relies on the "opinions" of other neurons in the network; in the Hopfield network, each neuron solicits the opinion of every other neuron before making up its mind. Recall that every neuron is a binary device and can render only two opinions: 0 or 1, off or on, no or yes.

Like our student, a neuron doesn't give all incoming opinions the same weight; it will listen to some neurons more than others. The degree to which the output of one neuron influences that of another is defined, once again, by the connection weight. Suppose the threshold for a given neuron is 3; for that neuron to turn on, it must receive a total of three "yes" opinions from other neurons. This sum could be achieved in countless ways—for example, as three yes votes from neurons connected by a weight of 1.0, or thirty yes votes from neurons connected by a weight of 0.1, or three hundred yes votes from neurons connected by a weight of 0.01, and so on.

Biological and electrical networks are quite large, containing millions or even billions of neurons. The number of interconnections can be equally huge; a human neuron connects to approximately 10,000 companion neurons. Therefore, even the very weakest "opinions" may influence a neuron's behavior when considered in the aggregate. The same holds for physical systems. A magnetic crystal is only trivially influenced by distant crystals, but the number of distant crystals is so gigantic that their cumulative effect can be significant.

Let's now build a sample network and see how it behaves. To make our network more understandable, we'll again express it in human terms, using a student analogy like the one employed by William Allman in his excellent book *Apprentices of Wonder: Inside the Neural Network Revolution*:

One thousand high school students are recruited and taken to an auditorium containing a thousand desks. Each desk is connected to every other desk by a wire, and on each wire is a resistor of variable size. (Resistors are devices that impede the flow of electric current.) Each desk also comes equipped with a six-volt battery, a light switch, and an electric current meter that registers from 0 to 100. The meter measures the total electric current flowing into a desk from all other desks. The light switch regulates the flow of electricity from each desk's battery to the other desks in the network. A desk is a bi-stable device: it must be either completely on or completely off.

The students are seated and given these simple instructions:

1. You aren't permitted to interact verbally with other students.
2. You can do schoolwork or read comic books, so long as you look at the electrical meter every minute or so and see where the needle is pointing.
3. If the needle on the meter rises to 50 or greater, turn your switch on; if less than 50, turn it off.

That's it. We have a network. Each desk becomes a "neuron."

The meter performs the additive function of a McCulloch-Pitts neuron, telling the student how much total electricity flows from the other students' batteries into his or her desk. The student performs the decision-making function, flipping the switch on when the needle crosses the threshold of 50 and switching it off when the needle falls below 50. The resistors serve as connection weights, limiting the amount of current that can flow over each separate wire. A large resistor permits little current flow between desks (a weak connection); a small resistor permits great current flow between desks (a strong connection). For now, we'll assume that every wire has been fitted with a different resistor chosen randomly from a big box. The resistors are also fixed; they can't change value over time.

The students are told to be seated and to begin looking at their needles. Switches soon flip on and off throughout the auditorium at a constant rate. The network is, at first, "far from equilibrium," and meters bounce all over the place. After a few minutes, however, the teacher overseeing this project does something strange. She approaches a select row of ten students and flips their switches into a predetermined pattern (where 1 is on and 0 is off):

1 1 0 0 1 0 1 0 1 1

The teacher informs these ten students that they aren't playing the meter game for now. They are to leave their switches in the positions she has put them in, regardless of what their meters say. In electrical engineering lingo, she had *clamped* these ten neurons—their switches are fixed, and they can no longer respond to input. So what happens to the other 990 students? An amazing thing. Switches start flipping very rapidly for a few minutes, and then—silence. The switches no longer move.

We just saw an E surface in action. After the ten students were clamped, the network quickly rolled into an equilibrium valley and came to rest at the lowest point. To understand why this happened, imagine that the electricity flowing around the auditorium is like water running over a wavy sheet of waterproof fabric. The contour of the network's electrical "fabric" is largely determined by the hundreds of randomly assigned resistors; the resistors selectively direct current flow over the various wires. Just as water droplets on the fabric must flow along a path of least resistance, electrons also must flow through the network along the easiest possible path. The hundreds of resistors help to create an uneven electrical landscape for the current to navigate. The students also contribute to the uneven electrical terrain by acting as tiny valves, turning the current flow on and off at various checkpoints.

We now see why this is called an energy surface. Water and electricity choose the path of least resistance, because that path requires the least expenditure of energy. Water doesn't flow uphill if it can flow downhill, and it won't flow down a gentle slope if it can barrel down a steep one. Similarly, electrons find the path of least resistance through any circuit. In our student network, the E surface isn't just theoretical; the network must seek its lowest state of electrical energy, and it must do so by navigating a geometric landscape of resistors and switches. Because the resistors were chosen at random, the shape of our student landscape has also been randomly generated. We've simply tossed the electrical fabric onto the floor to create a series of random bumps and creases.

When we clamp a subgroup of students, we are, in effect, initializing a drop of electricity somewhere on that random surface. (The students we clamp, and the pattern in which we clamp them, determines where on the surface we place this drop.) After it falls on the surface, the drop barrels at electronic speed through the landscape of resistors until the whole network achieves a minimum energy state and comes to rest in the closest attractor basin. After the ten students are clamped, the remaining students watch their needles as the electricity in the network seeks the optimal routes through the resistors. Eventually, the least favorable pathways are all switched off and the optimized pattern of current flow becomes fixed.

At that point, the network comes to rest and all switching activity ceases. The drop of "water" in this case is a binary string, a tiny piece of information. This information forces the network into an energy valley, so that the information and the valley are now linked. The network is a purely physical device obeying physical laws, but it's beginning to show the first signs of intelligent behavior.

After the network activity ceases, the teacher asks the remaining 990 students to record the position of their switches at that point in time. This provides a graphic record of the steady-state configuration induced by the clamping of ten students in a certain sequence. After the students finish marking their positions, she promptly unclamps the first ten students and the network starts roaming aimlessly again. Following an interval of time, she asks those 990 students to return their switches to the marked positions, the positions they were in when the ten students were clamped, and leave them there regardless of needle position. She has now clamped the 990 students into the previous steady-state position. But she left the original ten students unclamped. So what happens to them?

Things are now reversed. The original ten are now the only students free to follow their meters in the whole network, and they start flipping their switches for a brief time until they, too, halt. The whole network has reached a steady state again. The teacher now records the position of these ten students, and finds this pattern:

1 1 0 0 1 0 1 0 1 0

It differs from her original pattern by only a single digit. The steady state defined by the switch positions of 990 students has forced the original ten to revert to their clamped positions to a 90 percent accuracy. A string of binary data has been stored, within a margin of error, in the energy valley.

The teacher again unclamps all students, lets them roam for a bit, and then initializes the network again, this time with an *incomplete* version of the original ten-digit string. She clamps only seven of the ten students, like so:

1 1 0 0 1 0 1

The network again comes to rest in the same steady state as it did when all ten were clamped. When the system finally comes to rest and all the switching stops, she finds that the final positions of all ten students are once again:

1 1 0 0 1 0 1 0 1 1

After initializing the system *close* to an attractor basin, by using an incomplete set of data, the network rolls into the same valley and completes the three remaining digits to reconstitute the original pattern of ten. The teacher has "degraded" her original input of ten digits by reducing it to seven, but the network doesn't seem to care. It responds the same way it did when initialized with ten dig-its. In this model, we're already seeing the first inkling of how E-surface memory works, and how it leads to pattern completion. But wait, there's more.

The teacher randomly picks twenty students and asks them to leave their desks and eat lunch. (Assume that the ten students that were first clamped are not among the students asked to leave.) She repeats the previous two experiments, and discovers to her sur-prise that she gets the same results. Now she has "damaged" her network, knocking out 2 percent of its neurons, but it still func-tions. Try knocking out 2 percent of the transistors in a PC and see how well it performs. We see now why scientists from the vacuum-tube era showed an early interest in network machines.

The student network works well with degraded input, and it works well even when its own structure is damaged by the removal of neurons. Serial computers, on the other hand, can't tolerate ei-ther degraded information or damaged equipment. For example, if I have a file named *incometaxes* stored on my PC, I can't call it up with the file name *incometax*, even though the names differ by only two letters and both mean the same thing linguistically.

Unfortunately, biological systems must deal with degraded input all the time; the real world never feeds us perfect data. I use the word *degraded* here in a technical sense to mean any alteration of a data set. A pattern doesn't have to be damaged in an aesthetic sense to be degraded from its original form. If my father grows a beard, his image has been "degraded," yet he still looks great, and I'll still know who he is. (Lois Lane and Jimmy Olsen must have had serial processors for brains, since they could never recognize Superman after he put on a pair of glasses.) The real world is a dangerous place; living computers must be relatively impervious to injury, and networks provide the necessary degree of robustness. A person can lose large portions of his or her brain and still function normally. The same architecture that allows networks to complete patterns also buffers them against neuronal loss.

Our student network appears very promising. Unfortunately, however, it isn't, at least not in its present form. This particular network won't behave at all as I've described, not without a little additional modification. It has a flaw: the choice of connection weights. We've assigned the resistors *randomly* and made them incapable of changing over time. A random network really knows nothing and, what's worse, it can't learn anything new. The teacher's network, like any intelligent device, can't effectively recall strings of digits until it has been *taught* to do so—and the teacher hasn't taught her network anything yet. But all is not lost; far from it. We can make this network function exactly as I've described—much better, in fact—by simply adding that one, final ingredient: learning. I left "learning" out of the student model initially in order to make the core ideas simpler to understand. Now that we have the basics, we can embellish our model further.

E surfaces created by random resistors look like the Rocky Mountains, not like a smooth piece of fabric. Our random student E surface will have flat regions and knifelike peaks. If we place a drop of electrical information on a flat spot, it won't go anywhere, and if we put it on a sharp peak, the drop won't roll the same way twice. The essence of memory is reproducibility, and our random network has no reproducibility. Our goal is to use the E surface to store useful information by correlating that information with valleys in the surface. To do that, the network has to have control over the E surface and be able to change it with experience.

The connection weights define the E surface and so, if the network is to learn anything, the weights must change if the surface is to adapt to store new information. *In network theory, learning is synonymous with alterations in the connection weight matrix.*

Voilà! Four chapters into the book and we've hit pay dirt. We've finally arrived at the greatest Yang-Um boundary of them all, the true interface between the dead and the living, between the dumb and the smart. The critical distinction between a physical network, like a piece of magnetic rock, and an intelligent network, like the cytoplasm or the brain, can now be stated succinctly: *living, intelligent networks modify their connection weights in response to external training.* Magnetic crystals will interact in the same manner until the end of time, but the interactions within living networks evolve continuously. The connections among my brain's neurons at this point in time...

...aren't the ones I have now. The very act of writing this brief sentence has forever altered the social dynamics among the cells inside my head. But this begs the million-dollar question: just how does a network accomplish this feat? How does it translate external experiences into internal-connection weight changes? That all depends on the network. For now, we'll consider the simplest form of network learning: Hebbian learning.

In 1949, D. O. Hebb postulated that neurons in the brain follow a simple rule: whenever two neurons are "on" simultaneously, their connection weight will slowly increase over time. Otherwise, it will either stay static or weaken. (Technically, the original version of Hebb's rule only strengthened connections; we now know that connections must be both strengthened and weakened during Hebbian learning. We'll revisit this point in later chapters.) Hebb postulated that memories are imprinted in neural circuits as the most frequently used connections are reinforced during repetitious training. He was right, although decades would pass before the neurological basis of Hebb's astounding prophecy was discovered.

Hebb's rule asserts that chronically active neurons will strengthen their mutual bonds, forming little cliques, or social clubs, within the brain. As the network learns a pattern of external data, these cliques will coalesce to form an internal representation, or map, of that pattern. A Hebbian network organizes itself in a way that mirrors the outside world. Hopfield assumed that his networks used Hebb's rule, so let's return to our students and assemble a new network along these lines. When we first build the network, we still assign the initial weights randomly—the first E surface remains the unpredictable "tabula rasa" of our prior network. But we'll now install the Hebbian rule, and allow pairs of active students—students whose switch is "on" most of the time—to tweak their connections up a tad. Inactive students will see their connections weaken.

Along comes the teacher. Again she clamps the ten students in a specified pattern and then unclamps them. She clamps them a second time, in the same pattern, then unclamps again. She repeats this cycle over and over again. In the first student network, the random one, she was merely storing the ten-digit string in whatever valley happened to be nearest, and if, by chance, that valley was nice and smooth, the network might store and recall the information correctly. Most randomly created E surfaces, however, would fail to recall a ten-digit string accurately, so she decides to train the E surface to do a better job.

With the Hebbian rule in place, the teacher can now mold the E surface by repeatedly exposing it to the same information. Although a random E surface may prove ill suited to storing that particular drop of information, she can remodel that surface to suit her needs. Each time the teacher exposes the network to a certain "clamped" configuration, the Hebbian rule kicks in, strengthening some connections while weakening others. Her chosen drop of information begins carving a smooth pathway through what was once rugged, random terrain. In the first network, she dropped information onto the random terrain and simply took note of where it rolled. Now, she repeatedly hammers the terrain with droplets and lets them erode a unique pattern into the landscape, like a river carving out a canyon. She's creating an E surface custom-fitted to her information, creating new, smooth valleys to store new memories. This is Hebbian learning, and it enables us to imprint external data onto a landscape in a physical system, as Hopfield hoped. Learning, the act of changing the energy landscape in response to external training, becomes the missing link between the Yang of the dead and the Um of the living, the Yang of the dumb and the Um of the smart.

There are certain limits to this kind of learning, however. Suppose we train the network, through repeated clamping and unclamping, to recall the ten digits perfectly. The training process redefines the connection weights, via the Hebb rule, and remodels the E surface. That's fine. But what happens when we train the network to learn a different sequence of ten digits? This new piece of information will redefine the connection weights again, resulting in further alteration of the E surface. Each new experience changes the network's E surface to suit its own needs. Eventually, fresh experiences will begin to interfere with old ones. The surface is, after all, finite, and we can wedge only so many smooth valleys on it before they start overlapping. If the various memory valleys get too close, the network will start making mistakes.

In a Hopfield network, the number of memories the E surface can hold is proportional to the number of neurons. A student network that has 1,000 neurons, and that employs simple Hebbian learning, can reliably store about 130 independent memories before the valleys begin overlapping—according to the most recent research, the number of non-overlapping attractors a Hopfield network can store is approximately 13 percent of the total number of its neurons. Thus, size does matter here; bigger networks have a greater capacity for data. That's why an amoeba can store more information than a bacterium, and why my brain can store more information than an earthworm's.

Interestingly, new valleys don't pop up anywhere on the E surface. The more similar the information two valleys store, the closer they will be on the E surface. Very large E surfaces will be organized geographically according to the types of information they contain. Consequently, the confusion that arises from an overloaded network may not be all that bad. As valleys overlap, the network will make mistakes, but these won't be bad mistakes, given that overlapping valleys store similar information. For example, my overloaded brain may mistake "A Hard Day's Night" for "She Loves Me," but I'll still know they're both Beatles songs. In the "Beatles" region of my brain's E surface, the tunes overlap, making me prone to mistakes. Nevertheless, I'll never mistake the Beatles for Aaron Copland.

The geographic layout of network E surfaces endows them with a property called content-addressable memory, or CAM. In computer terms, an "address" defines where in the machine a memory is stored. In serial machines, like my PC, addresses are complex numbers that point the way to registers in a memory chip. Unfortunately, this numerical type of address provides us with no information about what's contained within it.

In the real world, however, an address *does* have meaning. For example, an address on the Upper East Side of Manhattan, say,

Sutton Place, will give us some general idea of who lives there, since the inhabitants of that area share certain common characteristics. So, too, the "address" of a memory inside a network has meaning, or *content*. When I speak of a "Beatles" region of my brain's E surface, I'm being very literal. Networks organize memories according to the content of those memories. Once again, this all ties in with the idea of pattern completion. Suppose I want to look up a certain Chinese proverb but I can't quite remember what it is. I do recall that it has something to do with fishing. If my local library were organized like a serial computer, every Chinese proverb would be assigned a number. My proverb might be number 88238967500. All I would have to do to find it is look up that number. Chances are, however, that if I can't remember the proverb, I can't remember a numerical address either.

Fortunately, libraries, like networks, have content-addressable memories. In other words, things are catalogued according to their content. I can remember two things about my proverb: it's from China and it's about fishing. In this case, my Chinese proverb "lives" in a book called, amazingly enough, *Chinese Proverbs*. Notice that the proverb's "address" inside the library—the book's title, in this case—says a great deal about it, just as an address on Sutton Place says a lot about a person who lives there. The book happens to be on a shelf containing only books about Chinese literature.

I turn to the index and find a proverb about fishing, and eureka! It points me to:

"Give a man a fish and he eats for a day; teach a man to fish and he eats for a lifetime."

This is the precise proverb I had in mind.

I was able to locate a specific proverb in a vast sea of information using only two pieces of content: Chinese proverbs and fishing. The library acted like a vast E surface, with data addressed geographically according to content. I navigated the three-dimensional space of the library guided only by the specific content "China" and "fishing." The library then enabled me to complete an incomplete piece of information. Had the library been organized like my PC's memory, I wouldn't have stood a chance of finding the proverb. That's why serial computers won't tolerate even the slightest degree of data degradation. If I want to retrieve a piece of information, I have to know the exact address. Coming close won't do. These simple networks, aided by a few simple rules, are beginning to look like pretty amazing things.

To summarize once more:

Information in a network is stored as an internal energy landscape defined by the network's connection weights. The connection weights are continuously molded by the network's training and experience. Because the weights determine the pattern of information flow within a network, they define a set of internal "rules" that determine how the network as a whole behaves. As the network learns, the weights change and the network develops new sets of internal rules. This is how a network "makes up" its own rules in response to training, namely, by altering its internal connectivity as it gains more experience.

Here's yet another difference between serial and network devices: while the rules used by serial devices make sense to us intuitively (because we write them), the rules used by networks have meaning only to the network. The internal logic of networks, although effective in its own way, appears tortured and bizarre to us. We wouldn't be able to decipher how the student network uses a given set of connection weights to store a set of ten-digit strings, for example, because the weights simply look like random nonsense to us.

For this reason, if we examine only the isolated "connections" between, say, two enzyme pathways in a cell, or between two bacterial species, we will be at a loss to decipher how a cellular or bacterial network uses such connections to solve problems on a larger scale. Our lack of any intuitive "feel" for the idiosyncratic logic of living networks has made it difficult for us to appreciate their true genius. When we plot out the enzyme pathways in a cell, the cytoplasm looks chaotic and without design or purpose. That's because we're used to thinking of intelligent machines as being built from arithmetic functions and logic operators, not from connection weights.

Living networks are patterned after inorganic networks (and, in all probability, arose from them). Because a network's intellectual capacity is a function of the plasticity of its connection weights, inorganic networks, hampered by their limited range of connectivity, have an equally limited capacity for intelligence. The advent of malleable ("trainable") connection weights endowed life with a broader capacity for intelligence.

In a literal sense, molecular life is the simplest manifestation of organic intelligence, namely, a biochemical network. Some "scientific creationists" argue that life is irreducibly complex and hence must be the product of Divine Intervention. What utter nonsense! In the network paradigm, the progression from inorganic networks (like spin glasses) to organic networks is, at least theoretically, seamless. The Hopfield network is basically a spin glass that learns. Life can be viewed in much the same way.

One final thought before continuing on to the next chapter. Notice that the networks we've examined so far are built using only "local" rules. Neurons and connections don't know or care a lick about what the network at large is doing. The McCulloch-Pitts neuron, for example, is a selfish little beast that pays no heed to anything but itself. Similarly, the Hebbian rule modifies connections on the basis of what's happening to two neurons, not according to what's happening in the network at large or in the outside world. The Hebbian rule itself—strengthening connections between active neurons, weakening connections between inactive neurons—has a strongly Darwinian aspect, as we'll examine in chapter 6. Networks that use only local rules to modify their connection weights can also be *self-organizing* in that they arrange their own structure without external help.

Step back and begin to see the grand scheme. Hordes of selfish little beings—socializing among themselves—unite to form a machine with emergent cognitive properties, a vast Darwinian engine bound by the same rules as physical systems. All living systems own the self-organizing capacity for emergent intelligence; that's what makes them alive. By congealing into social cliques, nonlinear things like enzymes, lymphocytes, and neurons can unwittingly mold themselves into great landscape maps capable of assimilating and manipulating environmental data. In the great creationism debate, two sides emerge—those who advocate evolution and those who advocate for intelligent design—but this divide is false. Evolution and intelligence are one and the same process.

I should note here that the simple Hopfield network, as described in this chapter, has severe limitations, both in terms of the types of problems it can solve and the accuracy of those solutions. In AI research, Hopfield's embryonic creations have great theoretical and historical value, but they can't be used as literal blueprints for making efficient machines. Living networks no doubt employ the principles first discovered by Hopfield and other connectionist

pioneers, including David Hillis, Geoffrey Hinton, Terrence Sejnowski, James McClelland, David Rumelhart, and others too numerous to name. The precise architecture of any biological
network, whether a cytoplasm or a brain, remains unknown, but it
is sure to be infinitely more complicated than any of our present
network models.

The Hopfield network provides the missing link between the
world of the dead—magnetic materials and spin glasses—and
the world of the living. The Leonardo quote at the beginning of
this chapter has, at long last, been fulfilled. Living systems are, at
their core, still bound by the laws of the physical world.

5 NETWORKS

*Traveler, there is no path
By walking, the paths are made.*

—Antonio Machado, *Campos de Castilla*, 1912

Hopfield showed that networks could be useful after all, and his work marked the beginning of a new era in network theory. Connectionist researchers soon discovered other types of networks and began applying them to practical problems with considerable success. The theory of parallel computation has come a long way since the days of the first two-layered perceptrons, but luckily, we don't need a comprehensive understanding of all networks here—most of them have no application to living systems. For our purposes, we need only to understand how a generic network operates. When boiled down to their bare bones, living networks probably use the same basic rules that we applied to the Hopfield model in the last chapter. I'll now recap what we know and flesh out some further details. (Keep in mind, we're not talking specifically about computers or computer science here, but about a general paradigm of computation that transcends any particular device, living or dead. It just so happens that networks are the paradigm of choice for living machines.)

To build a basic network, we need just two things: neurons and connections. Let's examine each separately. First, consider the neuron. To the anatomist, the word *neuron* evokes images of the giant, electrically active nerve cells that inhabit animal brains, but in the network sense, a neuron can be any nonlinear input/output device capable of forming connections with its neighbors. Mechanical relays, vacuum tubes, transistors, nerve cells, enzyme-catalyzed chemical reactions, bacteria, lymphocytes, and even human beings can all serve as neurons in some form of cognitive network. Of course, we mistakenly cling to the belief that brains are the paragon of organic intelligence, so we still refer to network building blocks as neurons even when they aren't true nerve cells. In the same vein, researchers sometimes use the phrase *neural network* to describe any connectionist device, although few networks are neural in the anatomic sense. Ironically, animal nerve cells were among the last forms of computational neuron to evolve (the first being the enzyme-catalyzed reaction).

The McCulloch-Pitts neuron is the simplest model of a neuron, little more than a modified on/off switch. Real neurons can exhibit far more complicated input/output behavior, although living systems may contain devices very similar to the McCulloch-Pitts device. In network theory, the quantitative relationship between a neuron's total input and its final output is called the *activation rule*. The condition of a neuron at any one point in time—on or off—is known as its activation state. (Some authors refer to network components as activation units, or simply as units, but this terminology strikes me as entirely too mundane—we'll stick with neuron.)

In our student network, the activation rule was rudimentary: when the input was 50 or greater, the student/neuron turned the switch on; when the input fell below 50, the student/neuron turned the switch off. Often, this simple on/off rule (also called a step function, because it resembles a step) proves sufficient, although nothing prevents us from making up more complicated activation rules.

For example, one of the most commonly used activation rules in artificial networks is the *sigmoidal rule.* This rule softens the abrupt transition from off to on by interposing a sloping region between the two states. Graphs of a neuron's activation state as a function of total input for the on/off rule (the step function used by the McCulloch-Pitts device) and the sigmoidal rule appear below:

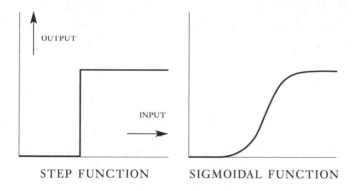

STEP FUNCTION   SIGMOIDAL FUNCTION

Mathematically, both graphs belong to a common family of neuronal activation rules; the step function on the left turns out to be a special case of the more general sigmoidal curve on the right. The sigmoidal rule is also widely used in living systems.

The sigmoidal (meaning "S-shaped") curve blurs the on/off transition, changing it from sharp to "fuzzy." The sigmoidal rule resembles the Tae Geuk of the Korean flag and embodies the same principle: the boundary between opposites is, in the real world, rarely distinct. Sigmoidal activation rules introduce noise into the neuron by creating some uncertainty about its activation state—is the neuron on, is it off, or is it trapped somewhere in between? In a world devoid of noise, the transition between two states is as sharp as a knife edge and everything becomes either black or white because, without noise, there can be no shades of gray.

Noise plays a large role in computer network models, and it figures prominently in all biological networks, too, as will be seen. In living systems, noise is a resource, not a liability. In fact, genetic noise drives the evolutionary process itself. It's impossible to come up with any new or original ideas in a black-or-white world devoid of uncertainty, since creativity lurks in life's gray areas. But for our present purposes, we need only remember that a neuron's activation rule can be simple or complex, as dictated by the needs of the network.

We turn now to connections. A connection can be defined as any weighted channel through which information flows. Here again, brain chauvinism influences us by forcing us to think of all connections as wires, either the actual wires inside a computer or the wirelike axons inside a nervous system. We mistakenly view neurons as immobile creatures, forced to relay their messages across fixed telephone lines, but in the real world, neurons can be highly mobile and capable of forming connections with other mobile neurons face-to-face, "on the fly," and without the need for rigid, permanent wires.

This all seems a tad confusing, so an analogy is in order:

Imagine a large party at a major corporate headquarters. Let's assume that I'm invited and that I know only a few people there. I enter the room and begin smiling vacuously at the many unfamiliar faces. Suddenly, I recognize a casual friend standing far across the room and we immediately head toward one another, politely weaving through a sea of strangers. We meet and begin exchanging small talk at close range, but, because we really aren't close friends, the conversation soon dies and we then head off in search of more fruitful interplay elsewhere. Later, I recognize a very dear friend and again we make our way toward one another to spend the remainder of the evening in a secluded corner, arguing politics.

Although I was "connected" to almost everyone in the room (I could at least smile and make eye contact with them), these connections were extraordinarily weak, so I formed and broke them quickly. I had a somewhat stronger connection to my casual friend—we exchanged pleasantries for a few minutes. The strongest connection by far was to my good friend. Note that the people in the room aren't hardwired like transistors on a circuit board; nevertheless, weighted information constantly travels between selected pairs of mobile partygoers, and even a group of mobile neurons eventually will achieve some form of stable structure. The party is a human spin glass that ultimately cools into an intricate mosaic pattern of like-minded "neurons."

From a theoretical point of view, the connections among mobile party guests are entirely equivalent to those in a hardwired network. In a hardwired network, neurons sit still and send bursts of information to other neurons via wires. At a cocktail party, the neurons (people) move about in search of other neurons, then exchange information through some short-lived form of communication that operates strictly at close range (in this case, conversational speech). In both cases, information passes between pairs of uniquely connected neurons.

But how does a mobile neuron discriminate among the many other neurons it encounters? A hardwired neuron can communicate only with those neurons to which it is wired, but a mobile neuron can bump into any other neuron in the network. In the case of the mobile neuron, the specificity (and weight) of the connection must be defined by some form of mutual recognition that occurs between connected neurons. At the corporate party, this recognition took place visually, through the mutual identification of familiar faces.

The weights in a mobile network behave just like the weights in a hardwired network and serve the same purpose: to define a unique pattern of network connectivity that can be exploited to store and retrieve information. At a party, some pairs of guests are highly connected, others are weakly connected, and still others are not connected at all. As in any network, these weights change over time: if two guests fight, they may not speak again, thereby severing a once strong connection entirely, or two strangers may create a new friendship and strengthen a previously weak connection. These connections have a strong Hebbian flavor; people with similar views will strengthen their mutual connections and gravitate into cliques. During the course of an evening, the social landscape will evolve into a form of content-addressable memory, with different viewpoints arranged in different areas of the room. Conservatives? They're over there, by the punch bowl.

In a "party network," mutual recognition among mobile neurons determines both the connections and the connection weights. Mobility enables neurons to seek out other neurons in a crowd, but without some form of mutual recognition, no true network can emerge among mobile neurons. A collection of water molecules can't form much of a network, because water molecules interact indiscriminately. Some authors have grandiosely proclaimed that any collection of things—from quarks to stars—can form a thinking network, a cosmic brain. This is foolish. The intelligence of a network derives from a unique connectivity that can be altered as the network learns, and quarks and stars don't manifest the fine-grained, unique connectivity needed to generate true intelligence. Party networks require a massive degree of neuronal individuality, something quarks and stars lack, because this individuality replaces the need for fixed wires. The entire universe can be built from a handful of different quarks, whereas it takes a million different lymphocyte species to make a thinking immune system. We see now why organic "spin glasses" are so much more intelligent than metallic ones—the wide variability in organic structures affords living networks a far greater range of connectivity than is possible within inorganic ensembles. The interactions among proteins, for example, are far more diverse than the interactions among gold and iron crystals in an inorganic spin glass.

If we outfitted our party guests with earplugs, mouth gags, and blindfolds, they would simply bump into one another like so many water molecules, gold atoms, or quarks. Stripped of their individuality, no social structure would emerge. When we remove the plugs, gags, and blindfolds, a social pattern quickly takes shape as little cliques form, dissolve, then gather again elsewhere. A cocktail party may look amorphous, but it still possesses an intrinsic structure defined by the relationships—the connections—within it. This structure contains information that spontaneously emerges in any ensemble of uniquely connected individuals. The individuals have no global agenda, and no central authority tells them what to do. By searching out old friends, or making new ones who share their views, they self-organize into a complex network.

Is a detailed discussion of networking in corporate parties really all that important? In fact, it's *extremely* important. Many living systems (the cytoplasm, the microbial mind, the immune intellect, an ant colony) consist of mobile neurons connected into party networks. Recall that enzyme molecules, hormone receptors, and antibodies come equipped with unique binding sites that recognize only a narrow range of targets. These binding sites serve the same function as the faces of my partygoers, allowing weighted connections to emerge in a vast soup of free-floating neurons. Binding sites enable enzymes in a cytoplasm (or lymphocytes in an immune system) to find friends in a large crowd. Enzymes and lymphocytes diffuse through their respective domains searching for familiar "faces" and, once in close enough proximity, they can exchange information with selected friends at close range using a chemical language analogous to conversational speech. The binding sites of organic macromolecules are infinitely rarer than human fingerprints, and this immense individuality endows life's mobile networks with a connectivity that can't be achieved by inorganic molecules, a connectivity vital both to being alive and to being intelligent. *It is the complexity inherent in carbon-based polymers that distinguishes the cytoplasm from its dim-witted precursor, the inorganic spin glass.*

In molecular and cellular networks, the strength of the interaction between chemical binding sites and their targets, known to chemists as *affinity,* defines the connection weights. The closer the three-dimensional "fit" between a binding site and its target, the higher their mutual affinity and the stronger their molecular connection. To put it another way, my foot has a high affinity for a size 11 men's shoe and a very low affinity for a women's size 5 high heel.

Recall also that during repeated exposures to antigens, certain connections within the immune system strengthen and others weaken. Chapter 2 described how antibodies can, through the genetic evolution of their binding sites, increase their affinities for foreign targets over time. One of the hallmarks of a mature immune response is the production of antibodies with an extraordinarily high affinity for foreign antigens. Thus, the immune system has the capacity to change its connectivity in response to external training. Sound familiar? It should—that's how networks learn, by altering their connection weights as they gain experience with the outside world. Exposure to foreign antigens alters the connection weights of the entire immune system, reshaping the immune E surface and installing new memories in the network's energy landscape.

Immune learning is similar to the Hebbian learning discussed in the last chapter: constant interaction among active components yields a higher affinity of interaction, while infrequent interaction leads to lower affinities. Once the immune system has been trained to recognize a pattern, subsequent infections become exercises in "attractor recall." Even incomplete fragments of antigen will send the immune system quickly into a trained valley, like a handful of notes sending my brain into a "pop tune" valley.

The bacteria of chapter 1 also form a party network. In the case of bacteria, the information exchanged is often genetic, and the connection weight between bacterial cells becomes proportional to their likelihood of trading genes. In this regard, some microbial species are poorly connected and others highly connected. Viruses help mediate bacterial connectivity; in a sense, *the bacterial connection weight matrix is under viral control.* Bacteria and viruses exist in a computational symbiosis, combining to form a more rapidly learning network. We can make the more general statement that parasites mediate connectivity throughout the biosphere; like it or not, humans, like bacteria, are connected by viruses—as well as by mosquitoes and a host of other parasitic "vectors." Although we consider parasites to be menaces, they are part of life's global informational engine. The microbial mind would be lost without them.

The "wires" linking enzymes in a cell, bacteria in a colony, or lymphocytes in an immune system aren't wires in the computer/ brain sense; rather, they resemble the connections at my cocktail party. Nevertheless, those connections still function like wires. Years ago, I first drew attention to collections of mobile neurons, calling them "fluid networks" (Ricard Sole and others have since written extensively about such networks). The name "party network" now seems preferable, given that mobile human beings and ants can make networks without floating about in some fluid phase.

Unlike human parties, collections of lymphocytes and enzymatic reactions have a strong statistical component. In human parties, it makes sense to consider the interaction between two individuals, but in molecular and cellular networks, it makes more sense to consider bulk (statistical) behavior. Thus, a neuron inside a cellular party network may experience not a *single* enzyme-catalyzed reaction but *huge numbers* of identical reactions occurring throughout the cell. Likewise, an entire clone of genetically related lymphocytes may act as a single neuron in an immune network.

The connection between two party neurons becomes a statistical property, representing the *average* affinity of two large groups for each another. The statistical connection weight resembles a cross-section in particle physics; the cross-section represents the *statistical probability* that two particles will collide. In a party network, the connection weight can be defined as the statistical cross-section between two populations, as defined by the frequency with which their components interact. Consider a party composed of university faculty members from four departments: physics, mathematics, philosophy, and English literature. Each department is equally represented. Suppose we keep track of all significant conversations (say, any that are more than five minutes long) between faculty members of two different departments. During an evening, we find the following:

Conversations between physics and math professors 38
Conversations between physics and philosophy professors 14
Conversations between physics and literature professors 6

We can infer that the interaction between the physics and math professors had a higher statistical "cross-section" relative to the interaction between physics and literature professors, and that makes sense. Physicists, as a statistical group, have a higher affinity for mathematicians than they do for experts on *Beowulf.* In giant statistical ensembles comprised of identical units—enzymes, lymphocytes, and (as will later be seen) ants—connection weights are more accurately defined by studying the *bulk* behavior of interacting groups, rather than studying the behavior of individuals.

Inside the brain, axons and dendrites, the filamentous projections arising from nerve cells, create neural connections. Nerve cells can't move, and so they're forced to make their connections through wires. They also trade their individuality for a unique circuitry—nerve cells all look alike, or nearly so. Ironically, this form of connection, the one we consider most typical, was actually among the *last* to evolve, just as the nerve cell was the last type of neuron to appear. Prior to the evolution of nerve cells, all living networks were of the party variety, comprised of moving neurons. So why did wires evolve at all?

The appearance of hardwired connections allowed larger and more complex networks to arise. Party networks work well at the molecular level because molecules travel extremely fast, but such networks lose steam as their size increases. The computational speed of party networks depends upon the statistical mixing and diffusion of mobile neurons. On an atomic scale, mixing and diffusion occur at a phenomenally quick pace—an enzyme can bump into countless billions of other proteins in a nanosecond. Although I consider myself a fast mover, I can't roam a party that fast. Large party networks made up of mobile neurons will "think" only as fast as their neurons can move, and neurons get progressively slower as they, and their networks, grow larger.

A single enzyme can jog the length of a cell in milliseconds (or faster); consequently, enzyme networks work quickly, even without wires. (The cytoplasm has fixed structures that may serve as wires, but that's beyond the scope of our present discussion.) A nerve cell, on the other hand—even one that can move as quickly as a pond amoeba—can't make the trip across my head in less than an hour. Imagine how ponderously I would move if my nerve cells had to travel from my brain to deliver messages personally to my thigh muscles.

Neurons can't move, but they can do something far better: they can send a message across the brain, via one of their axons, in a millisecond. In other words, the brain, a hardwired network tipping the scales at over two pounds, can work as fast as a chemical party network operating on a molecular scale. Wires do for biological networks what mass communications do for human networks. The explosion of living network technology that occurred when axons and dendrites evolved (the advent of brains) was replicated in the explosion of human technology that followed the invention of the telephone. After nerve cells evolved, cells no longer had to crawl large distances to exchange information. Simple physics tells us that wires became a speedier alternative to the party paradigm as neuronal size increased and neuronal mobility decreased. Eventually, neuronal mobility was abandoned altogether in favor of a hardwired format. Large nerve cells could now send information over fixed connections far faster than they could deliver it personally. The evolution of the axon—biology's wire—opened a new era in computational speed among cellular networks. The dense wiring inside a human brain also allows a single neuron to interact with thousands of others simultaneously, a degree of connectivity beyond the scope of a mobile network. Brains took the basic network concept—the concept that defines life itself—and refined it by hardwiring the connections. The individuality of wires replaced the individuality of macromolecules.

Brains are a frozen version of the life process itself, although mobile networks were the first to evolve and they remain the prototype for biological computation. The unique interactions between a mobile enzyme and its substrate, or between two friends at a party, don't mimic a wire; *the wire mimics them.* The wire evolved not as the first connection, but as an *alternative* to the far more ancient free-floating connections among mobile neurons. The hardwired nature of the brain creates the illusion that evolving nervous systems abandoned the realm of network computation in favor of rule-based or "digital" computation, but that's all it is: an illusion. A brain thinks like a cytoplasm, not like a laptop.

So why do some networks, like the immune system or a colony of ants, still use the party format? It's all a matter of the best design for the best problem. I don't need a Formula One race car to drive my daughter to school. Likewise, immune systems and ant colonies don't need to solve problems as quickly as brains do. I have less than a second to jump out of the way of an automobile, but up to two weeks to respond to an influenza virus. Moreover, the molecular nature of information an immune system must handle—toxins, bacteria, and viruses—forces it to retain a party architecture. In contrast, the brain receives information chiefly in the form of sound and light, which can easily be digitized into a hardwired input. The issue of necessary computational speed is an important one. Sometimes the speediest network isn't the best network; slow networks may be just right for dealing with slowly evolving problems. Ants and other eusocial insects form party networks that reason more slowly than an immune system, given that the bugs move even slower than lymphocytes. Nevertheless, these networks work well, because the problems they face don't require rapid solutions.

The failure of both mainstream biology and theoretical AI to recognize the importance of party networks has become a significant barrier to creating a unified model of biological intelligence in all of its manifestations, from cell to ecosystem. Brain chauvinism has been overwhelming in this regard: to the average AI expert, only hardwired things can truly think. Nevertheless, I challenge any network theorist to examine a large social gathering of people (or termites, for that matter) and fail to see network principles in action. Even in a seemingly random mob, an E surface takes shape and the mob adapts, as one entity, to its environment.

Let's take a deep breath and summarize what we've covered.

A network is made from neurons and weighted connections. Neurons can be any input/output device (enzymes, cells, organisms); the relationship between a neuron's input and output is defined by its activation rule.

Neurons communicate via weighted connections. In party networks, like the cytoplasm and bacterial communes, mobile neurons move about and exchange information directly with other neurons. The specificity and weights of "party connections" are often defined by the statistical affinity of one mobile neuron (or group of neurons) for another neuron (or group of neurons). In hardwired networks, like the brain, neurons sacrifice their mobility and instead exchange information over fixed wires. Theoretically, party networks and hardwired networks work in much the same way, but from a practical point of view, there are some differences. Party networks work best on the very small scale (where neurons can move quickly and molecular-sized wires are less reliable), while hardwired networks work better on the large scale (where neuronal slowness makes the party format less efficient). There are exceptions, however. Ant colonies and human societies (large-scale networks) employ a party format, while some cells (small-scale networks) may employ some form of cytoplasmic wiring.

Before moving on to the next chapter, we should probe a little deeper into the differences between serial computation and network computation, where I use the word *computation* to mean any process used to solve a given problem. We tend to equate computation with arithmetic, but it actually means much more than that.

Serial computers work best for problems that involve numbers and symbols. The reason for this is simple: serial computers rely on arithmetic and the formal rules of logic. They also like exact and reproducible solutions. When I use my calculator—a serial computer—I expect an answer accurate to ten places each and every time. Networks, on the other hand, deal best with *patterned information* and prefer *good* solutions to *exact* ones.

It's now time for another analogy:

Robert the Robot Maker wants to build a machine that can hit a decent golf drive. At first, he plans to use a serial computer for his robot's brain. Unfortunately, he's now forced to program it, because serial computers do only what they're told; and because serial computers work only with logic and arithmetic, Robert must figure out some way of breaking down a golf drive into a series of mathematical equations. This can be done, of course, but not very easily. Robert relies on a local physics professor to derive a complex set of differential equations that describes the perfect golf swing. The elaborate equations take into account the wind, the type of club used, and even the humidity of the air. The result is a robot capable of hitting a ball to a defined target within a six-inch margin of error. Well, at least in theory. There's just one problem: Robert estimates that the final program will take him two years to write. Even worse, his budget allows for only a modest computer, one that would take hours to make the necessary calculations for hitting a single ball. He gives up on this approach.

He tries a network computer instead. Now he doesn't have to program the computer at all; he can sit back and watch the network discover how to swing a golf club all by itself. How? By building a network capable of modifying its connection weights as it learns. Just as the student Hopfield network memorized a pattern of digits, Robert's robot can memorize a pattern of motorized movements that will produce a reasonable golf swing. Instead of the students' simple Hebbian learning, however, he builds a network that employs a more involved form of learning, known as the *back-propagation of errors*, a practical algorithm discovered by David Rummelhart, Geoffrey Hinton, and Ronald Williams in 1986.

Back-propagation of errors (called *backprop* by network mavens) forces the network to modify its connection weights in response to *output error*. Output error is the difference between the *actual* behavior of the network and the *ideal* behavior the network is striving to achieve. In backprop, the change in connection weights is proportional to the magnitude and direction of the output error. As the network gets closer to the ideal behavior, the corrections to the connection weights grow smaller.

Backprop resembles the process of hanging a painting on a wall. We toss the painting up haphazardly, then step back and judge if the painting is level. If it's tilting very far right, we walk over and make a large correction to the left. It may then be a little too far left—we've overcorrected—so we shift it back a small amount to the right. Now it's a tad too far right, but just a tad. We repeat this process, bouncing left and right in smaller increments, until the painting matches our preconceived notion of the ideal position on the wall.

Of course, we could have measured from the ceiling down and exactly calculated the ideal position of a level painting, which is what serial computers do. Serial computers like numbers and precision. Networks prefer quickly eyeballing things and making numerous corrections until they look right. This latter approach is much faster and doesn't require numbers, measurements, or calculations, but, in the end, it's not quite as accurate.

Instead of writing an elaborate program telling his robot what to do, Robert simply informs its network brain where he wants the ball to land after it is struck by the club: three hundred yards from the tee and straight down the middle. This becomes the ideal output, the standard the network will use for determining its output error. He then equips his robot with an optical system for measuring where the ball lands on the fairway.

Robert assigns the initial connection weights at random (essentially starting his robot out as a golfing nincompoop), props his virgin network onto the golf course, and lets it whack away. The first swing is, not surprisingly, a complete whiff—the network fails to make any contact with the ball. It compares this fiasco to the ideal output and realizes that it missed a three-hundred-yard drive by an astounding three hundred yards. (How I know that feeling!)

Clearly, it has made an enormous error. But instead of wrapping its driver around a tree, the network uses backprop to make large adjustments in the connection weights. It swings again, and this time the ball goes straight up in the air and plops five yards to the right of the tee. The network adjusts again. As the robot's drives get nearer to the ideal landing spot, the weight adjustments become smaller and smaller.

Eventually, the network gets to a point where it can't make any further improvements. No network is perfect, and Robert's robot is no exception. (Even when we think our painting looks level, formal measurements often show that it really isn't.) Robert tests his self-trained robot and finds that it can hit a drive fairly straight but strays from the target by an average of forty yards each time. This is not quite the six-inch margin of error his (unbuilt) serial machine would have achieved, but it's good enough for Robert's purposes. Even an ideal golfer doesn't need a six-inch accuracy off the tee.

For networks, learning how to hit a drive is an exercise in pattern recognition, not an exercise in physics or advanced mathematics. Our brains act like sophisticated versions of backprop networks (although whether they actually use anything resembling backprop is, as yet, unknown). We learn to golf in the same way Robert's robot does, through trial and error and making corrections as we go. I've learned what pattern of muscle movements will propel my golf ball in a certain direction, but I've never mapped out the physics of a ball's trajectory. Golfers don't care about the physics; they just know that when they swing like this, the ball goes there.

Programming a serial computer to do complex activities like hitting or catching a ball is cumbersome. These activities can't be easily broken down into a series of sequential mathematical and logical rules that can be digested, one at a time, by a microprocessor. Such activities are more easily learned as patterns. In fact, virtually every survival problem faced by living systems can be boiled down to a matter of pattern recognition: a tree sensing the changes in light and temperature that mark the onset of spring; a hawk scanning the brush for rodents; a white cell roaming the bloodstream in search of pathogens. These creatures don't calculate solutions the way serial computers do. If the tree were a serial computer, it would contain some instruction like "if daylight exceeds 15 hours in a 24-hour cycle and temperature exceeds 10 degrees centigrade for 10 consecutive days, then sprout leaves; else, remain dormant." But we know of no such programming in our backyard oaks. Life is a network, not a rule-based von Neumann machine, and networks adore patterned data.

Networks also loathe arithmetic. The human brain, more complex than the most sophisticated electronic computer, turns out to be a terrible calculator. Although a handful of people, including mathematical geniuses and some *savants*, can do amazing calculations without pencil or paper, the average brain has a terrible time doing even basic arithmetic. The greatest geniuses can't match a cheap calculator for accuracy and versatility. Some people can add two ten-digit numbers in their heads, but can't mentally compute the product of two three-digit numbers. Idiot savants— people of average or below-average intellectual ability who happen to be astoundingly good at one activity—can only do one form of calculation, and it's usually some oddball problem like taking the seventh root of any number below one million. The brain isn't built for arithmetic, nor is it built for following logical rules. We learn math and logic like we learn everything else, from golf swings to playing the trumpet: by learning and emulating patterned data, through experience and experimentation. Savants and math whizzes don't have a special computational part of their brains; they're just better at applying pattern completion to a peculiar set of arithmetic problems.

In the odd-couple world of intelligent machines, serial computers are the Felix Ungers and networks the Oscar Madisons. Serial machines demand neatness and precision at any price; networks will settle for less-than-perfect, even sloppy solutions that can be achieved quickly and cheaply. Felix wants to know why things happen; Oscar doesn't care why things happen, only that they happen in his favor. Both types of machine have their place. Serial machines belong in the offices of the Internal Revenue Service, where error is intolerable, arithmetic abounds, and cost is no object. Networks, however, are the machine of choice for living systems, whose bread-and-butter is pattern recognition. In the biological realm, Oscar Madison rules. Living things need the most efficient form of information processing for the job at hand, but not necessarily the best form. Exact solutions come at too high a price and are almost never necessary in the real world. The average person is a lousy golfer because the human brain didn't evolve to execute movements *perfectly each time.* Perfect execution is the province of serial computers, not living machines.

How does Robert's golfing network "see" the ball? And how does it run the motors that swing the club? In other words, how does information get in and out of a network?

Recall that a network neuron receives input and generates an output. So far, we've assumed that neurons talk mostly to other neurons, but that's not really the case. A neuron can also receive input from the environment and generate output back to the outside world. Remember the clamped students in the student net-work? These students served as the interface between the teacher, who represented the environment, and the network. When the teacher clamped these students, she was feeding a ten-digit string of data into the network from the external world.

Let's modify our student network. Rather than have our teacher manually clamp the switches of those ten students, we will instead outfit their ten desks with photoelectric cells wired into the current meters. Whenever a strong light strikes the cells, a current will flow through their meters, driving the needles above 50 and prompting the students to flip on their switches. The teacher can then clamp these students from across the room by shining a flashlight on their photoelectric cells. The ten desks will have become a very crude retina, converting light from the environment into electrical impulses that enter the network. (Note that the light-sensitive desks have become "connected" to the out-side world, and note too that these connections can be weighted by varying the efficiency of the photoelectric cells.)

We can modify the student network even more. Let's select seven other students and hook their light switches to seven electric tone generators, corresponding to the musical notes A through G. When one of these seven students switches his or her desk on, the corresponding musical note will sound. The teacher will know which of those students is on simply by listening for their corre-sponding notes. She can now play games, flashing her light on the photoelectric cells in a certain sequence and listening for the pat-tern of notes that sound from another region of the network. By modifying the connection weights of her network, she could, at least theoretically, create a network that plays simple melodies in response to specific light patterns. The melodies themselves would be stored in the E surface created by the whole of the network.

The modified student network now has three distinct classes of neurons: ten input neurons (the students with photoelectric cells, who take information from the environment and feed it into the network); seven output neurons (the students who produce musical tones, thereby taking information from the network and pumping it back into the environment); and 983 *hidden* neurons (the remaining students, who interact only with other students and have nothing to do with the outside world). All neurons still follow the McCulloch-Pitts rules, but now differ in how they gather input and distribute output.

Like the earliest multicellular creatures, our modified network now has the beginnings of cellular differentiation, with different neurons serving different roles.

Although the input and output neurons carry the fanciest hardware, the real business of the network takes place in the hidden neurons. They're called "hidden" because the environment has no direct effect on them and they have no direct effect on the environment. (For network purists, fully connected Hopfield networks don't possess hidden units in the strict sense, but I'm aiming for clarity here, not technical rigor. There's a lot in this book that will irk network purists, but some oversimplifications are necessary to keep the core ideas from being swamped in minutiae.) They are buried deep inside the machinery of the network. Hidden neurons manipulate, or "filter," environmental information as it passes from the input neurons to output neurons. In this way, the network's experience—as imprinted in the connection weights linking its hidden neurons—modifies how a given input produces a desired output. In the final analysis, all intelligent machines are basically just input/output devices. The magnitude of a machine's intelligence is reflected in how well inputs and outputs are correlated.

For a golfer, the input is a visual image of the fairway and the output is a golf swing that places the ball in the correct location. The hidden neurons, the brain, store the network's trained connection weights, weights that should convert a given input into the desired output. The golfer's skill—the degree of his or her golf intelligence, so to speak—is measured by how often he or she can convert any input into an ideal output. What we call "practice" is the act of honing the hidden network filter (through the Hebbian training of connection weights among hidden neurons) so that any possible input will always be matched with an ideal output. That, my friends, is network intelligence in a nutshell.

We now have the tools to see the fatal flaw of the first perceptrons: as it turns out, they had only input and output neurons, no hidden neurons. (Later versions equipped with hidden units, called multilayered perceptrons, would prove superior to their predecessors.) If brains were built like the original perceptrons, our eyeballs would be wired directly to our muscles, with no cerebral cortex intervening. The simplest creatures have brains something like this. If you shine a light on them, they move away—not very sophisticated behavior, to be sure. For an outfielder to catch a fly ball on a dead run, however, hidden neurons must come into play. The hidden neurons store the years of experience and training that a diving catch requires. Hidden neurons make the difference between simple reflexes and complex actions.

In the human brain, most neurons are hidden (99.9 percent of them, to be precise). These neurons create the internal landscapes that give us our inner voice and endow us with imagination. They are, quite literally, the engines of our dreams. Now we see why networks like patterned data: they can manipulate patterned data easily by recreating those patterns internally, in the form of connections among their hidden neurons. Data that enter the network must traverse the internal landscape of hidden networks before exiting the other side as output. As data filter through the hidden landscape, the network's E surface, they become transformed. In this way, a network's prior experience, now imprinted on that landscape, determines how a network responds to a given input. Networks "think" by using their internal landscapes as data filters between input and output neurons. As a ball approaches my hand, my brain passes the visual data from my eyes through the hidden landscape of my brain and into my muscles so that I might catch the ball. Practicing this act imprints the right pattern into my cortex and makes my internal landscape of hidden neurons a better filter for passing my sighting of the ball (input) into my intercepting of the ball with my hand (output). No formal calculations or logical rules are needed. Networks do employ a certain internal logic, but we would find that logic undecipherable.

The input neurons of the nervous system consist of specialized nerve cells like the rods and cones of the retina (vision), the hair cells of the ears (hearing), the sensory organs of the skin (touch), and so on. The primary output neuron of the nervous system is the myocyte, or muscle cell. Anatomically, muscle cells aren't true nerve cells, even though they're electrically active. Nevertheless, muscle cells are also McCulloch-Pitts neurons; they have two states: on (contracted) and off (relaxed). Moreover, like any McCulloch-Pitts neuron, the myocyte adds up incoming signals from the network (the nervous system in this instance) and decides whether that input crosses the threshold for switching states from on to off or vice versa.

The brain also signals glandular cells, which have a similar McCulloch-Pitts behavior: on (secrete hormone), off (do not-secrete hormone). The nervous system looks similar to our modified student network. It consists entirely of neurons. Although some neurons (rods, cones, myocytes, glandular cells) have been outfitted with elaborate hardware that enables them to interface with the outside world, the input/output cells still participate in the network at large. The brain itself represents the hidden neurons that control how a given input yields the correct output. Note that the input neurons of cytoplasmic networks, the membrane receptors that serve as the cell's eyes and ears, are simply modified enzymes; the output neurons are also enzymes. This illustrates a general rule of networks: their input and output devices are specialized neurons, but neurons nonetheless.

In the immune system, some cells recognize antigens (input cells) and other cells target the enemy (output cells). Even though much of the immune system becomes active during an infection, only a few types of cells deal directly with the outside world. Does the immune system also have hidden neurons—lymphocytes that have nothing to do with recognizing antigens directly but still participate in the cognitive process of immunity? Do hidden lymphocytes store internal representations of the antigen world, much like the cerebrum stores internal representations of the visual world? This is a tantalizing idea that has received little or no attention from mainstream immunologists.

We do know that the immune system manifests something akin to content-addressable memory. If we immunize an animal with a large protein, multiple antibodies form against different areas of that protein. Although the antibodies have completely different V-region binding sites, they share a common idiotype. I mentioned idiotypes back in chapter 2, but didn't dwell on them very much. Idiotypes are markers within or near the V-region of the antibody molecule; they act like little ID badges that can be recognized by other lymphocytes. The late Niels Jerne once suggested that idiotypic markers regulate the internal interactions among lymphocyte populations. If Jerne's idea proves right, then idiotypes help create the weighted connections among hidden lymphocytes.

The fact that different antibodies directed against a common antigen share the same idiotype suggests that the immune system uses idiotypes as keywords in a kind of librarylike filing system, a content-addressable form of storage and recall. Because they share a common idiotype, the diverse set of antibodies that form against a single antigen can be turned on or off simultaneously by a hidden lymphocyte that accesses that idiotype (an anti-idiotype lymphocyte). Anti-idiotype lymphocytes meet the criteria of hidden neurons: they regulate the flow of information but serve no direct input/output role.

If hidden lymphocytes do exist, the immune system has the capacity for associative intelligence. In its own way, it can think like a brain. Our bodies may then consist of two discrete minds, one operating at the macroscopic level and the other at the microscopic level. This is a rather creepy kind of Yang/Um dualism, but again, I digress.

One final technical note: Robert's robot illustrates the last ingredient of a true network: *the learning rule.* The learning rule defines how a network changes its connection weights in response to external training. Learning rules can be modified to make a network better at solving specific classes of problems. Some learning rules allow associative recall, others are better at classifying objects, and still others excel at feature extraction. In the last chapter, we explored the earliest and simplest learning rule, Hebb's rule. Robert's machine used backprop, a more modern and sophisticated type of learning rule.

These two learning rules also differ in a much more fundamental way. Hebb's rule is a *local* learning rule, because it affects only one connection at a time using local information. Hebbian learning states that two neurons alter their mutual connection on their own, without any external guidance. Backprop is a *global* learning rule, because it uses an external agency, or trainer, to change many connections simultaneously using information gleaned from the output of the entire network. Local and global learning rules are also referred to as "unsupervised" and "supervised" learning rules, respectively. As previously discussed, local learning can lead to "self-organization" as well. Networks that use local learning rules must learn in an unsupervised way, without an external tutor to guide them as they change their connection weights to model the external world. To appreciate the difference between local and global learning, let's return to our corporate party.

Most parties self-organize according to local learning rules. In other words, individuals decide with whom they are going to associate and how intense those associations will be. But what happens if the boss arrives and starts telling everybody what to do? For example, assume that the boss goes to the microphone and orders all people from marketing to assemble by the bar. The boss is upset because the party has become dull and lifeless. Based upon that information, which represents his impression of the network's output as a whole, the boss makes connectivity adjustments intended to liven things up. The marketing department does as it was told and, pretty soon, everyone in marketing starts talking and exchanging information. They may not be talking to the people they like best, but they have no choice. The boss has changed dozens of connection weights at the same time by stripping individuals of the right to make their own associations.

Solid-state versions of neural networks are now commercially available, and most of them use some variation of the backprop global learning rule. These networks have a "boss," a separate sub-program that calculates the difference between the network's real output and its ideal output and makes changes in the network's weights according to some error-correcting algorithm. The boss program, called the network trainer, forces large groups of neu-rons to change their connectivity at the same time, whether they would want to or not under the Hebbian paradigm. Global learn-ing is a very effective and easily implemented form of learning for hardwired simulations of neural networks, which is why Robert liked it for his transistorized Tiger Woods. But it's not feasible (or desirable) for most living networks.

Although advanced nervous systems might use some form of global learning, I think it's unlikely that party networks—cyto-plasmic networks, immune systems, insect colonies, ecosystems, global economies—could employ this approach. Party networks have no "boss" to tell them how to alter their connectivity, and so they rely on local learning rules and organize themselves locally as best they can. Besides, having a "boss" can defeat the whole pur-pose of the network, since a boss excessively centralizes the whole intellectual process. Networks exist to distribute computation over many components, not to entrust all learning to a benevolent dictator.

Living networks favor the local approach because living neu-rons are, well, selfish. They do what they want and prefer to make their own connections as suits their selfish needs. They "walk" where they please and, in doing so, wear down memory paths. It is this selfishness that leads us to a critical link, namely, the link between network behavior and Darwinian "survival of the fittest" evolution. That link becomes the subject of our next chapter.

6

THE SELFISH NEURON

Darwin's "survival of the fittest" is really a special case of a more general law of the survival of the stable.

—Richard Dawkins, *The Selfish Gene*, 1976

In his popular book *The Selfish Gene*, zoologist Richard Dawkins challenges the idea that evolution is driven by the survival of species. On the contrary, said Dawkins, evolution isn't about the survival of *species*, but about the survival of *individuals*, or, more precisely, about the survival of the *traits* manifested by those individuals—traits defined by genes.

The nineteenth-century version of Darwinism, as outlined by Darwin himself in his landmark treatise "The Origin of Species," gave way to neo-Darwinism in the latter half of the twentieth century. Neo-Darwinism fused Darwinian concepts with the theory of modern genetics, and, in this context, the "selfish gene" model represents neo-Darwinism taken to its extreme. By defining evolution as a fitness contest fought by genes, Dawkins sees the world not as a Galápagos paradise of competing finches and lizards, but as a seething cauldron of competing DNA.

I must disagree with this concept. Genes can't be selfish, or selfless for that matter. They are what they are: inert bits of information strung like precious pearls on strands of ribose sugars.

Such raw information can't propagate itself or pursue any other selfish motives. Evolution possesses an intrinsic intelligence, that's true, but that intelligence doesn't originate from static entities like genes. Although biological intelligence is an emergent property of many simpler individuals working in concert, those individuals responsible for evolution's emergent intellect can't be genes, because genes aren't sufficiently dynamic. Those who view genes as the irreducible building blocks of life are mistaken.

The basic building block of life is the neuron.

I use the term "neuron" here in the abstract, as it was used in the previous chapter. A neuron is any nonlinear input/output processing device operating within the context of a larger network. It can take many practical forms, as already observed—nerve cells, chemical reactions, bacteria, lymphocytes, even human beings. These various neurons may differ radically in their physical construction, but within their respective networks they all serve the same role. In the earliest days of the primordial soup, the first and only neurons would have been the multi-stable enzyme-catalyzed reactions we examined in so much detail back in chapter 3. Life began as a molecular network long before the existence of genes. So why did genes arise, and how do they fit into the network model of living systems?

Thanks to our gene-centric view of living systems, no theory of life's origins has been wholly satisfactory. We see life as an elaborate stage production written, produced, and directed by genes, but modern genes are clearly too complex to have arisen spontaneously. The ingenious design of the genetic code, or so the creationist argument goes, could only be the result of Divine Intervention. Only a Supreme Being could have engineered such a marvelous bit of cybernetic machinery out of whole cloth.

Our thinking in the matter of life's origins has been clouded by a genetic chauvinism, which is, on the small scale, entirely equivalent to brain chauvinism on the large scale. We can't imagine life without genes any more than we can imagine intelligence without brains, yet life *had* to have existed before DNA (and before the existence of the genetic code), unless we want to side with creationists and assume that modern life, genes and all, arose instantaneously by supernatural edict. Even if we accept that genes are the building blocks of life today, they couldn't have been the building blocks of life at the beginning of biological time. Life came first, genes later. The first living things had to rely on something other than DNA for data storage and transmission, but what?

Recall that DNA is simply another form of written language, a molecular hieroglyphics encoding the linear amino acid sequences of protein macromolecules. The story of life may be written in DNA now, but that story began long before the existence of genes. In human experience, we know that storytelling predates any written language. Our stories were once passed by word of mouth—a serviceable, albeit limited, mode of data transmission. The advent of our written languages may have increased the amount and accuracy of verbal information that could be stored and later retrieved, but it didn't change the fundamental nature of human discourse. The ideas in Homer's *Iliad* transcend the symbols used to encode them on paper for future generations.

Geneticists are akin to pure bibliophiles, enamored of books but not of literature. Life's magic resides in the information stored within genes, not in the genes themselves. DNA isn't some mystical substance fallen to earth like manna from heaven—it's the molecular equivalent of cheap newsprint. Molecular biologists, experts in DNA biochemistry, will strongly disagree with me on this point. They will argue that they are indeed interested in the genetic information DNA contains and will point to their recent successes in deciphering the complete nucleotide sequences of many organisms, including the human organism, but they can't explain how their genetic sequences create a complete multicellular being. In short, they have piecemeal knowledge about proteins and genes, yet show no grasp of the mechanism as a whole. They see the trees quite well, but don't comprehend the forest in the least.

Molecular geneticists now possess reams of raw data about many diverse life forms, including the human genome itself, but they still can't explain how a dynamic interplay among macromolecules allows a living system to arise from lifeless matter. In fact, recent revelations about the relatively small number of human genes has only deepened the mystery of how so few genes can cooperate to form such a complex being. Mainstream biologists gaze upon their DNA sequences as I would upon the schematic drawings of a Pentium III microprocessor—with a mixture of amazement and ignorance. The more we know about genes, the more bewildered we seem to become about the origins of life itself.

A massive research effort has been focused on the minutiae of genes with considerably less effort expended for understanding the system-wide behavior of the living things those genes create. If computer scientists were like biologists, they would know everything about the transistor and nothing about the way transistors cooperate to make circuits. The average biology graduate student can ramble on about leucine zippers, topoisomerases, introns, and other highly technical issues, yet would likely stumble over any attempt to define what life is or why it exists at all. This isn't mere speculation on my part. I recently addressed a gathering of senior medical students, most of whom had been biology majors as undergraduates, and asked them to define "life," the very process that they had spent much of their adult lives studying. No one even ventured to try. Formulating cohesive models of living systems, what I call life's cybernetics, has been relegated to fringe thinkers (including yours truly). To understand how living systems work, we must shed our genetic chauvinism and start looking at genes in a new way.

Back to the problem at hand—how do genes relate to networks? Let's revisit the student Hopfield network in which the connections follow Hebb's rule. With the Hebbian rule in place, the resistors between desks are no longer fixed but can now be modified according to Hebbian principles. Assume that the teacher trains her students to store a large set of ten-digit strings using Hebbian learning. As she repeatedly exposes the students to the data, they gradually alter their connectivity, with the most active students aligning into social cliques. When she's finished, the final set of trained connection weights will define an electrical E surface; the stored memories—the set of ten-digit strings—will be valleys, or energy minima, in that surface. If the teacher later prompts the network with a partial subset of the data, thereby initializing the network near one of these valleys, the network will roll the rest of the way into the valley and recall the remainder of the data. This form of attractor-based pattern memory should be quite familiar by now.

With the experiment concluded, the teacher disbands the students. The next day, however, the superintendent of schools asks her to give a demonstration of her trained network to a group of government officials interested in her work. Unfortunately, it had taken her all day to train those students and she'd never be able to locate all of them again so quickly; she can't possibly recreate the network on such short notice. Or can she?

Indeed she can. She doesn't need to locate the original students. The network's knowledge now resides in the connection weights, the variable resistors, not in one set of students. The act of training contours the weights and the weights, in turn, establish the basic geography of the trained E surface. So long as the janitorial staff hasn't disturbed the final settings of the resistors between desks, the teacher needs only to reseat any group of one thousand people and the network should function as it did when run by her original students. She rounds up a thousand volunteers, gives them some basic instruction, and the "new" network performs flawlessly.

The government officials are impressed. So impressed, in fact, that they ask the teacher to take her "student" network (which no longer contains students) to the state capitol and demonstrate it to the governor. But the act of transporting a thousand desks and people, not to mention countless resistors, meters, batteries, and wires, across the state would be far too costly. Still, she has an idea. Instead of transporting people and desks, she can rebuild her network much more cheaply elsewhere by hiring local people and renting local desks. Moreover, she doesn't have to retrain the new network—she only has to *write down* the final values of the resistors in her own trained network and set the new resistors to exactly the same values. (Easier said than done, given that even this relatively small Hopfield network has close to a million resistors, but the principle still holds.)

The teacher records all the connection weight values on a huge sheet of paper. That paper now contains the complete information needed to recreate her trained network. Of course, she will need workers and electricians to translate that written information back into a three-dimensional set of wired desks suitable for the new students at the state capitol. Nevertheless, the connection weights, which define the custom-contoured E surface of her original network, have been entirely reduced to a set of encoded symbols on a portable piece of tablet paper. She could rebuild the same trained network anywhere, at any time, using this piece of paper. So what does this have to do with genes?

It was noted in the last chapter that the cytoplasm is a form of party network wherein the connection between any two macro-molecules is defined by their mutual chemical affinity. In partic-ular, the binding affinity of proteins comes from their unique three-dimensional contour. The contour of a protein, as was ex-plained in chapter 2, arises from its amino acid sequence, which, in turn, is linearly encoded in the form of a gene. Thus, *the connec-tion weights among the proteins inside the cell are "written down" on a tablet of DNA, using the genetic code as language.* DNA serves the same role as the teacher's portable piece of paper, storing all of the con-nection weights needed to recreate a new version of a previously trained network somewhere else. In this case, the trained network is the cell itself. A cell's chromosomes contain the blueprint for its unique E surface.

Let's now assume that our teacher was illiterate—extraordi-narily unlikely, needless to say, but we can assume anything we want in thought experiments; that's what makes them so fun and useful. If she were unable to write down the connection weights, how would she demonstrate her trained network at the state capi-tol? She would then have only two options: transport the original network across the state, desks, resistors, students, and all—costly, but possible—or build an untrained network at the new location and retrain it again on site. The latter option seems more feasible. Remember, it took the teacher only a day to train the original net-work. She would need to arrive early and pay her local students for an extra day's work, but that would be the easier solution to her dilemma.

But what if the information stored inside the teacher's net-work's connections were more complex than simple ten-digit strings, such that it took her many months, not just one or two days, to train the network successfully? For example, suppose she built the network containing photoelectric cells and musical note generators and taught it to play several Broadway show tunes in response to light signals. Believe it or not, this type of pattern training is possible, even in such a simple network, but it certainly can't be done in a day. Without some means of recording the final connection weights in a portable format, the teacher's ability to recreate a laboriously trained network in new locations would be hampered by the need to reproduce the elaborate training process elsewhere. So, too, with living networks. Without some way of recording the connections among proteins in a portable format, complex living networks would find it difficult to create faithful copies of themselves.

The first "pre-genetic" protein networks could store only limited amounts of data, since they had to arise spontaneously in many different locales without the means of recording the connection weights among proteins in DNA shorthand. Consider the human case. Geographically isolated networks of prehistoric humans could solve simple problems locally, like inventing the wheel, without written language, perhaps even without spoken language. At this early stage, each human community had to solve the problem anew and, once that network dissipated, their solution would be lost forever. The advent of human language allowed old solutions to be recorded for posteriority, freeing up time and computational resources for solving more complex problems. When Henry Ford made his first Model T, he already knew about the wheel. The emergence of genes in protein networks, like the emergence of language in human networks, was a landmark event that greatly expanded the ability of networks to solve environmental dilemmas. Life existed before genes, just as human storytelling and technology existed before language, but the capabilities of life exploded after the development of the genetic code.

Like prehistoric human communities, pre-genetic networks of

proteins had to solve every problem from scratch since they lacked any written DNA record of prior solutions. Unfortunately, when a particular network expired, its architecture dissipated and the information it had learned soon vanished from the face of the earth. At this point, life was still possible, but its intellectual progress was extremely slow. The advent of genetic language gave living protein networks a way to store previously trained sets of connection weights in a permanent and portable way, and thus networks in different locations didn't have to "reinvent the wheel" each time, but could now build upon the training of prior networks.

The advent of genes also redefined the nature of network learning. Recall that, in network theory, *learning is the act of modifying connection weights in response to experience.* After the appearance of the genetic code, the connection weights among proteins were written in the genes. DNA now stores the "weight matrix" of living networks. Consequently, training a living protein network now means altering—training—the genes themselves. The alteration of genes in the natural world comes chiefly via mutation and other random errors in DNA handling. (To simplify things, I'll refer to any DNA "error" as a mutation, even though that's not technically true.) We now arrive at a critical postulate in the network model of life:

> *Genetic evolution is the way protein networks "learn" in response to experience, namely, through mutational modification of their connection weights at the DNA level.*

This seems rather odd. How do random events like mutations drive a nonrandom process like network learning? The answer is fairly straightforward: learning occurs by coupling random mutations to a *local competition* that favors certain types of mutations over others. In other words, we can easily inject Darwinism into a network by allowing only the "fittest" connection weight changes to survive. This form of natural selection takes place locally, on a small scale, yet generates network learning on a large scale.

Random corrections to network weights can work so long as we have some means of sorting out the good corrections from the bad during the learning process. In the backprop model of picture hanging, our corrections contain a random element. True, we know by looking at the painting how to correct it—shifting it left or right, to a large degree or small—but we never know how to make any given correction *exactly*. That's why we rock the painting back and forth multiple times until we get the picture level; we need time for our successive iterations to cancel out any random error. Electronic networks can be programmed to use exact learning rules that make nonrandom corrections to the weights as the network learns. In living networks, however, learning is the result of *random* changes in connectivity coupled to a Darwinian selection of the *best* changes.

Critics of Darwinian evolution, including the proponents of the quasi-religious "intelligent design" school, argue that natural selection alone can't generate the complexity we see in the modern biosphere. They're wrong—natural selection does generate complexity, *but it does so only by serving as a learning rule in a network paradigm.* Darwin and his intellectual descendants have extracted the learning rule of natural selection from the genetic network and placed it on a pedestal all by itself, ignoring the network architecture inherent in the biosphere as a whole. By viewing natural selection out of its network context, mainstream Darwinists make evolution seem far less intelligent than it really is. Survival of the fittest, like any local learning rule, can exert its power only within the machinery of the network. Darwin saw the rule, but not the device in which it operates.

As I mentioned several chapters ago, if we disassemble networks into their component parts, we will have great trouble deciphering how they work. Remember, networks (unlike serial computers) are emergent devices—they work their magic through cooperation and so don't yield their secrets easily upon dissection. Hebb's rule and backprop are, mathematically speaking, extraordinarily simple, yet when incorporated into electronic networks, these rules can enable networks to do some amazing things, from forecasting the movement of the stock market to defeating chess champions. If we looked at the backprop rule by itself, however, outside the context of the machine, we would be dumbfounded to explain how this simple rule could yield a chess-playing (or golfing) device. Likewise, Dawkins and his ilk are forever tap dancing around the fact that natural selection alone seems incapable of creating human beings. The point is, natural selection alone *can't* yield great complexity; only natural selection operating in a network paradigm can.

Most people are familiar with the little hand-held pinball games in which tiny metal balls are sealed in a clear plastic case. The goal of the game is to get the balls into the appropriate holes. At first, we try to steer the balls directly into the holes by tilting the case this way and that. When this proves impossible, as it almost always does (particularly if we have two or more balls to deal with simultaneously), we resort to agitating the game until the balls happen to drop into the right spot by sheer luck. Once the balls are in the holes, we stop shaking the case.

These games illustrate the power of random events coupled to some selection process. We shake the game until we get the right outcome, then we stop. Living systems are continuously agitated—by ionizing radiation, UV light, heat, physical trauma, and so on—and they exploit this randomness to achieve a directed goal. The first way of solving the pinball game—precisely steering all balls directly into the holes—is the approach serial computers would take. The second approach, relying on random agitation and selecting for the right outcome, is the computational scheme of choice for living networks. The fact that we are good at the random way of locating the little balls (and not so good at steering them into the holes precisely) reflects the network organization of our brains and the Darwinian way in which we learn tasks.

One additional point about the random agitation of connection weights in living networks: because the weights are continuously agitated, some random shifting of the landscape takes place within living networks all the time, *even when they aren't being exposed to external stimuli.* The E surface has a certain degree of instability, like the ocean's surface. Nevertheless, random connection weight changes still must obey internal selection pressures, even when no training is occurring. A living network can refine its internal data maps in the absence of external training. In the brain, the agitation of connection weights creates a process known as thought. Even when isolated from the environment, the brain alters its connectivity and comes up with new internal correlations among previously learned data sets.

When viewed under a microscope, a speck of sand floating in water will bob and weave crazily, buffeted by the motion of the surrounding water molecules. The molecules in any substance vibrate, provided that substance has a temperature above −273° C (the so-called absolute zero point at which all atomic motion ceases); the higher the temperature, the more vigorous the vibration. Objects suspended in a sea of vibrating molecules will be buffeted to and fro by these vibrations, constantly moved by the thermal "noise" inherent in any physical system. This buffeting, called Brownian motion, was first explained by Einstein in 1905.

As it turns out, networks are also subject to background noise very similar to the thermal noise in physical systems. Even "at rest," networks can be buffeted from between stables states, bouncing about the E surface like a grain of sand dancing in water. *Thus, we can consider "thinking" as a form of Brownian motion in a network buffeted by internal random noise.* This phenomenon is not limited to brains, strangely enough. Technically, any living network, including a single cell, should be capable of something resembling thought. The speed and complexity of that thought may not approach that achieved by the human brain, but it's thought nonetheless.

In a Hebbian network, the "fittest" corrections are those that enhance the connectivity of active neurons. In a Darwinian network, the best corrections are those that reward a neuron for altering connections in the desired way, the reward coming in the form of greater efficiency, more food, or a better chance to survive. *If two active neurons are rewarded for strengthening their connection, then a Hebbian network becomes a Darwinian one*—more on this in a moment. In the case of Hebbian and Darwinian networks, corrections to the connection weights are applied and tested locally, at the level of individual neurons.

Neurons, not genes, become the selfish creatures here. Neurons are designed to be selfish. Just think about how they behave: (1) they respond only to their local input, with no concern for what's happening to other neurons; and (2) when using Hebbian, Darwinian, or other forms of local learning, they alter their connectivity to suit their own needs alone, not the needs of the network. In network theory, the word *local* becomes synonymous with Dawkins' more inflammatory term *selfish*. Amazingly, global network learning emerges from local (selfish) neurons employing self-centered learning rules. To observe this principle in action, we need to revisit the human immune system.

Recall that during an immune reaction to a foreign antigen, lymphocyte V-regions undergo random mutational changes. Some mutant V-regions have a strong affinity to the foreign antigens, and the lymphocytes equipped with these V-regions are stimulated to divide. Other mutant V-regions express lower affinity for the antigens, and their lymphocytes languish. The net result of this process is the dominant production of mutant V-regions with progressively higher affinity—a stronger connection weight—for foreign antigens. Because antigens act like food, or a growth stimulant, for lymphocytes, those lymphocytes with favorable mutations will bind more antigens and win out over those with unfavorable mutations.

By coupling antigen recognition to a system for rewarding lymphocytes, the immune system can select among random V-region mutations to produce a stronger connection between itself and a foreign antigen. The more frequent and prolonged the antigen exposure, the higher the affinity (connection).The accelerated genetic evolution of V-regions resembles Hebbian learning, in that frequent interaction between antigens and the immune system results in high-affinity antibodies and rare exposure produces lower-affinity antibodies. Clearly, the immune system is also Darwinian. Observe the link between Hebb's rule, a property originally inferred by studying brain networks, and textbook Darwinism.

The immune system provides us with a perfect laboratory for exploring how Darwinian evolution—the local (selfish) selection of the fittest random mutations over the less fit ones—can be used to train the connection weights within a cognitive network. In certain ensembles, like the immune system and bacterial colonies, networks seem aware of the link between random mutational change and connection weight training; these networks permit,

even encourage, a heightened degree of genetic instability to accelerate their learning.

Organisms "learn" via mutational changes in the connection
weight matrix encoded in their genes—that's what conventional
Darwinian evolution is all about. Immune systems and microbial
colonies, on the other hand, are unique in that they rely on a form
of mutational learning that has been intentionally accelerated to
take place over time scales ranging from a few weeks to several
years. Nervous systems, as will be seen in the next chapter, replace real mutations with simulated ones, accelerating the process
even more. Although the mechanics and time scales may differ, *all
biological learning, whether in the simplest cell or the smartest brain, is a
local fitness competition operating on the connection weights of networks.*
We can go one step further and define evolution generically as
that process that trains the connection weights in living networks,
thereby allowing them to acquire and recall patterned sets of environmental data. Note that this broad definition doesn't mention
genes at all.

Mainstream neo-Darwinism views evolution as an entirely genetic process, driven by mutations or other genetic aberrations,
but the problem with this paradigm is that it can't explain how the
genetic machinery itself evolved. In fact, it makes no sense at all
to speak of DNA "evolving," since modern evolutionary thinking
has been couched entirely in genetic terms. Herein lies the Gor-
dian knot of mainstream evolutionary theory: how could the genetic code evolve, when evolution itself has been defined as a
genetic process? Dawkins believes that the first biological entity
was a self-replicating molecule that he named the "replicator," and
that all subsequent life derived from this proto-gene. He doesn't
concede that there ever was a pre-genetic period of life. I find this
idea unpalatable, a flawed attempt to rescue the gene-centric view

of life at all costs. Although I don't doubt that self-replicating macromolecules existed in the pre-biotic soup, I wonder how they would continue to evolve. In the network paradigm, we can postulate not only a pre-genetic version of life but a pre-genetic form of evolution as well, one driven by the direct alteration of connection weights among proteins and other macromolecules within primitive enzyme networks. DNA isn't the only macromolecule subject to random modifications by free radicals, ultraviolet radiation, and other environmental toxins. In pre-genetic networks, random changes in connection weights would have occurred directly among a network's components—proteins and rudimentary nucleic acid polymers—as a result of structural damage inflicted by the early earth's hostile environment. (In post-genetic networks, direct network learning was later replaced by learning solely at the genome level.) Random pre-genetic alterations would still have been subject to natural selection, in that networks with "better" alterations would preferentially survive. Some form of pre-genetic evolution must have spawned the genetic code, ultimately transferring all responsibility for connection weight training within protein networks to the genes alone.

Interestingly, as will be seen in the next chapter, *nervous systems employ a form of post-genetic evolution, wherein the connection weight changes are again implemented in the network directly, without the need for permanent alterations in genetic molecules.* With the arrival of brains, evolution came full circle. Evolution is a learning process, an alteration of connectivity within networks in response to environmental training. This learning first took place non-genetically in primitive macromolecular collectives, then genetically, and, after the evolution of the nervous system, non-genetically once more. (Ironically, the advent of written language represents a shift back to genetic learning among human networks.)

Different systems use evolution differently, but the basic format remains the same: (1) generate random changes in connection weights at a certain rate, and then (2) allow local selection forces to choose the fittest changes according to some learning rule. These weight changes contour the network's E surface in response to environmental training, spawning a social organization among the network's neurons that harbors the internal representations of externally derived information. All that remains are mere details—what constitutes the neurons, how the weights are defined, how data enter and leave the network, the network's computational speed, the specific learning paradigm, and so on. These details separate a cell from a brain from an immune system from a rainforest. Differences aside, living networks think, learn, and evolve according to a shared set of network principles, the biologist's answer to the physicist's unified field theory.

The nature and rate of random weight fluctuations vary widely in different biological systems. In network theory, the rate at which the weights change is called the *learning rate*. In pregenetic networks, changes occurred directly in the macromolecular makeup of the network. Consequently, their learning rate was likely quite slow. In genetic networks, the learning rate equals the mutation rate, either insufferably slow (as in geological evolution) or not quite so slow (as in the enhanced evolution of immune systems and microbial colonies). In post-genetic networks (brains), the learning rate is extraordinarily speedy. We know from mathematical network models that the learning rate can affect the stability of the network. In some networks, rapid changes in connection weights can produce oscillations that prevent the network from settling into a steady-state solution. If we try to level a painting by making only large, rapid corrections, the painting will swing wildly back and forth without ever reaching its proper position. In simpler networks—cells, ant hills, bacterial colonies—this stability problem likely limits their learning rates severely. The nervous system, on the other hand, somehow manages to overcome this problem, at least partially.

Unfortunately, pre-genetic and post-genetic networks share a common weakness. Because their connectivity is trained directly at the network level, once they dissipate, their experiences and training dissipate with them. The first networks lacked permanence, just as my brain's experiences and thoughts lack permanence. The evolution of genes (and, much later, the evolution of human language) gave networks the means to immortalize their experiences.

In evolutionary theory, biological systems don't consider all possible states to be equally favorable. They prefer the state of "alive," as opposed to the state of "dead," since that's what survival of the fittest is all about. But in a system devoid of any learning properties, neurons have no preferred state. They can be either on or off; it really doesn't matter to them. Once we introduce some form of local learning behavior, like Hebb's rule, we force neurons to seek some preferred state at the expense of others. We make neurons selfish by committing them to seek that preferred state at all costs, without regard to their competitors' needs. The word *self-ish* makes sense only in the context of a group. If I'm alone in a room with a large cake, I can eat all I want without being selfish. I am selfish only if my consumption of cake impacts others who would also like to eat it. To be selfish, an entity has to (1) want something, and (2) belong to a group of others who want the same thing.

The simplest form of Hebb's rule, the one we've been using, can be stated as follows: if a neuron activates another neuron repeatedly, it will become progressively more efficient at activating that neuron. Thus, two neurons that stimulate each other to stay "on" will become more strongly connected over time. Hebb's rule can be extended further: a neuron that repeatedly prevents, or inhibits, another neuron from turning on will have its connection to that neuron weakened. If two neurons have little effect on each other, their connection may slowly decay over time, like a muscle atrophying from disuse. In the Hebbian paradigm, the preferred state of the neuron, its version of aliveness, is "on" and, like a living thing, it will seek allies that help it stay on (alive) and actively dissociate itself from enemies that seek to kill it by turning it off. The neuron gradually comes to ignore neutral neurons that neither help nor threaten it.

Hebb's rule and its many variations imbue the network with a fine-grained Darwinism; each neuron competes to stay alive within the ecological confines of the network. Like all finite things, a network possesses a limited resource of whatever substance flows through its myriad connections—electricity, chemical reactants, information, neurotransmitters, whatever. Hebbian learning pits neuron against neuron in a fitness contest for the common network resource.

In the student network, each desk requires a certain current flow to keep the needle above 50, thereby keeping the student "on" (alive). Assume that I'm one of those students. A friendly desk, one that frequently feeds me current, is worthy of a stronger connection. Any desk that saps electricity away from me could drive my needle into the dead range, and thus is a mortal enemy to be avoided. Although we really haven't designed the student network to have predatory desks that suck electricity away from other desks, it's possible to do so. Ergo, I am selfish. I want enough current to keep me alive, even if it means killing other desks or crashing the whole network. I don't care a whit about anything except keeping my needle above 50 and I'll make or break connections to that end, without regard to the global consequences of my greed. Networks that learn in this fashion are like miniature ecosystems, filled with competing neurons all fighting to stay alive. Since the vast majority of neurons are dependent upon other neurons for their sustenance, neurons must either form alliances or die. Some succeed; others perish. The checkerboard of neuronal alliances, viewed against the backdrop of their enfeebled or dead companions, creates a unique pattern similar to the patterns of sporadically aligned pockets of spins in a spin glass. This checkerboard carries the network's complement of learned information. *The network's experience becomes imprinted in the pattern of alliances made among hidden neurons in their competition for a shared network resource.* In living networks, global learning flows from local Darwinism. In the fitness contest among multicellular creatures, being "dead" is irreversible, although in some Darwinian networks, like the brain, it is reversible.

To be selfish, an entity must crave some limited resource in order to perpetuate its preferred state and be willing to compete with others to obtain it, regardless of the consequences to society at large. For evolution to flow from selfishness, a common resource has to be limited enough to make any competition for it mortal, so that only the fittest are left standing (although this mortality may be fleeting). Therein lie the problems with Dawkins' selfish gene model. Why do genes want to stay alive? Why would they care? Isolated molecules like DNA are like neurons outside of a learning network: they have no preferred state. The concept of a preferred state matters only in the context of the complete, dynamic system, and against what do genes compete? In the absence of a comprehensive network model, simply postulating that genes are selfish makes little sense.

An enzymatically driven reaction, on the other hand, can be selfish. First, a reaction depends on a local supply of reactants, a limited resource. Second, those reactants may be coveted by other reactions in the immediate vicinity, setting the stage for competition. Third, a reaction can have multiple stable states, including a preferred state as defined by enzyme kinetics. It makes *some* sense to think of a reaction as alive, or dead, in that it either is actively taking place or is not. Finally, enzymes have unique three-dimensional structures that make them capable of forming unique alliances with other proteins. None of these observations applies to isolated pieces of nucleic acid. Again, genes were *not* life's first building blocks, nor are they the most fundamental unit of living systems today.

Equating the Hebbian rule with natural selection may be a little bit of a stretch, I'll admit, but that doesn't matter. The important point here is that *natural selection acts as a local learning rule in earth's genetic network.* Once we understand this, we can easily see how the "simple" rule of survival of the fittest can yield the great miracles around us without having to invoke a Divine or extraterrestrial intelligence. The biosphere is, in the final analysis, of purely earthly design.

Before moving on to the brain, I would like briefly to consider one further topic in evolutionary science: *punctuated equilibrium.* In the late 1970s, Stephen Jay Gould and Niles Eldredge challenged conventional scientific wisdom by pointing out that evolution isn't the smooth, steady process Darwin envisioned, but a process marred by sporadic fits and starts. The fossil record suggests that the number and variety of species remain remarkably constant for long periods of geological time—in a sort of "species equilibrium." These equilibrium periods were occasionally punctuated by brief bursts of rapid change as multiple abrupt extinctions occurred, coupled to the rapid appearance of many new species. Several authors (notably Stuart Kauffman, a leading expert in complexity theory) have pointed out that punctuated equilibrium resembles a ball rolling over a "species fitness" landscape: the global ecosystem becomes trapped for long periods in a stable valley until some perturbation flings it into a new one. The transition to a new steady state (a punctuated equilibrium) wreaks havoc, causing a massive genetic upheaval worldwide. Punctuated equilibrium remains controversial, but if true, it is best explained in terms of network theory.

Of course, we don't have to invoke punctuated equilibrium to find evidence of the abrupt transitions suggestive of a ball roaming an uneven fitness landscape. The very existence of discrete species implies that some form of E surface governs evolution. A major criticism of Darwinism concerns its failure to explain how separate species arise at all. Although Darwin referred to evolution as the "origin of species," his theory is ill equipped to explain anything except how major variations arise within any one species, such as why finches evolved with different beaks or butterflies with different colorations. Darwin assumed that a smooth series of morphological alterations, over time, would eventually cause one species to transform into a different species altogether, but this linear model implies that a large number of intermediate forms must bridge two related species. The fossil record simply doesn't support this. Even Darwin was aware of this inconsistency. In the human lineage, for example, the vast morphological differences between apes and humans would mandate hundreds, perhaps thousands, of intermediary species to bridge the gap smoothly, but the requisite large numbers of missing links just aren't there. The jump from ape to human, like all evolutionary progressions between species, was accomplished by a handful of abrupt leaps, not by a parcel of baby steps. Long before Eldredge and Gould expounded their theory, evidence of punctuated equilibrium could be found in the solitary lineages of any modern creature.

Networks like to stay in their stable states. As we saw in the case of cancer, for a network to jump out of a stable state requires many things going wrong at once. Carcinogenesis—the act of dislodging a single cell from one stable state and jumping it to another state, or rolling it off the fitness landscape altogether—requires at least three critical errors in network behavior. Jumping an entire organism to another stable state—to another species altogether—would require even more simultaneous changes. By virtue of their design, networks don't permit a small number of changes to spark the jump to a completely new state; that's why minor transition forms rarely exist between species. Intermediate states between stable states are, by definition, unstable and fleeting.

We see the same phenomenon reflected in our own bodies. A cancer's lineage contains few "intermediary" forms. For example, a prostate cell turns into a benign adenoma (the cause of prostatic urinary retention), then into a frankly lethal carcinoma. Despite the vast differences among these three states, there appear to be no clear intermediate forms separating them, no smooth progression from benevolent cell to insane miscreant. The three stages—normal cell, benign tumor, malignant tumor—represent three unique species of cell and their progression, one to another, a microscopic recapitulation of the disjointed equilibrium "jumps" found in global punctuated equilibrium and hominid evolution.

But I've digressed long enough. On to the brain.

PART III
THE LARGE

7

WIDER THAN
THE SKY

The Brain—is wider than the Sky—
For—put them side by side—
The one the other will contain
With ease—and You—beside.

—Emily Dickinson,
"The brain is wider than the sky"
(poem 632) 1862

Many years ago, a young woman named Alice was brought to my emergency room. Earlier that morning, she had tripped as she dashed downstairs to check on a cake burning in her oven. We never found out what caused her to fall, but we knew that she had fallen a long way before coming to an abrupt rest, forehead-first, on a polished hardwood landing. She arrived in a coma, barely breathing.

A computerized brain scan showed that her right frontal bone had imploded inward and was lacerating the underlying brain. A thick hematoma now filled the lobe and threatened to smash the life out of the remainder of Alice's gelatinous brain. I would have none of that, so I whisked her to the operating room, where I unceremoniously sucked out the clot and lopped off the anterior half of the right frontal lobe. The lobe had been pulped beyond repair and would only cause her future grief—in the form of seizures, swelling, or further hemorrhage.

After a few weeks in the hospital and several months in a local rehabilitation center, Alice fully recovered and went back to

homemaking. She now lacks almost one-half of one frontal lobe—
a 25 percent deficit in her most critical cognitive hardware—yet
no one can tell the difference between the old Alice and the post-
lobectomy Alice, not even her husband. Her sense of humor, her
memory, her crocheting skills, even her atrociously bad bowling
game—all were completely preserved. How can this be?

We once believed that humans use only 10 percent of their
brains and so losing a random chunk of gray matter shouldn't hurt
us, but this is myth. Although we've been given more nerve cells
than we need to survive, just as we've been endowed with more
liver and kidney cells than are absolutely necessary, all of us use
most of our brains, most of the time. So we have a paradox here:
how can a person utilize nearly all of his or her intact brain, yet
lose a significant portion of it to trauma and emerge unscathed?

We could invoke plasticity, the brain's ability to rewire itself
after injury, but Alice's recovery was far too rapid for that. Her
cognitive skills recovered quickly after her brain surgery; her prolonged stay in rehabilitation was a result of a concurrent orthope-
dic injury (she had broken her hip in the staircase fall, too). Alice
recovered so rapidly, in fact, that she behaved as though she hadn't
been truly damaged in the first place. Something weird is at work
here. Do we need all of our brain matter, or don't we? The answer
to this puzzle, once again, lies in network theory.

Think back to our student network. When we randomly
deleted a certain percentage of students, the network still performed well. Networks recognize external patterns by mapping
them onto internal patterns formed by neurons and their connection weights during the learning process. Networks recognize
degraded patterns thanks to their ability to roll into E surface valleys, and they can also complete their own internal patterns after
an injury in much the same way.

When we first learn a new song, we must hear all of the notes. Once that song has been drubbed into our memory, individual notes—even whole groups of notes—become dispensable during the recall process. I've heard Bing Crosby's "White Christmas" so many times that I can mentally replay the entire song after hearing only a few notes. I need to hear the whole song to learn it, but not to recall it later. Similarly, once the brain's hidden neurons have been extensively trained to recognize a pattern or perform some task, many of them become dispensable to the recall process. It doesn't matter if the song's external pattern is degraded by the loss of notes or if the song's internal pattern is degraded by the loss of neurons; the recall process still works. A network isn't infallible, of course. A song can become so garbled, or a brain so diseased, that errors become inevitable. Nevertheless, the network's design allows the margin of error needed for data processing in an imperfect (and dangerous) world. When networks fail, they do so gracefully, in contrast to the catastrophic failures manifested by serial computers.

Whole slabs of nervous tissue can be lost without harming long-term memory; the more highly trained we've been to remember a fact or a task, the harder it becomes to eradicate that fact or task from our brains through physical trauma. The common phrase "it's just like riding a bicycle" turns out to be quite true. Once we learn a complex task well, we rarely forget it totally. To forget our first names, the most ingrained pattern in the brain, we have to be near cerebral death. The television version of amnesia—forgetting who we are—is creative fiction. Bilingual patients who sustain strokes in their speech areas or develop Alzheimer's disease will lose their ability to speak the most recently acquired language before they lose their native tongue. Demented patients may not recall events of the previous week, yet can easily remember the address of their first home. The brain is very good at completing trained patterns imprinted deeply into its connections, even after injury. This behavior comes naturally to networks, but not to serial, rule-based computers. Clinical neurosurgery provides us with our first indication that the brain is not a pure von Neumann machine.

Does network architecture make an intelligent machine immune to failure? No. Any physical device can fail; it's not a matter of *whether* a device will fail, but *how* it will fail. For example, a certain class of networks, called *scale-free networks,* are extraordinarily resistant to *random* neuronal failures but quite vulnerable to the failure of a few *critical* neurons. A scale-free network is the Achilles' heel of the network world—almost invulnerable, but not quite. (The word *scale-free* derives from the mathematical analysis of how these networks are connected and thus lies outside the scope of this discussion. Nevertheless, scale-free networks are more than a technical oddity, since all living networks likely have a "scale-free" architecture.)

Scale-free networks display an inhomogeneous pattern of connectivity—a neuron may be connected to two companions or to two thousand. In contrast, the student Hopfield network was homogeneously connected: all students had the same number of connections. Inhomogeneous connectivity makes some neurons more "important" than others. In fact, the average neuron in a scale-free design becomes expendable, making network behavior dependent upon a handful of elite neurons, and these neurons represent the Achilles' heel of the network.

Homogeneous networks are still quite resistant to injury when compared to serial machines like a personal computer, but they're not nearly as robust as scale-free networks. We can remove a few percent of students from the homogeneous Hopfield network and still obtain reasonable results, but we can't perform a network "lobectomy" and remove a quarter of its neurons without reducing the remaining ensemble to incoherence. Since living networks, from bacterial cells to the Internet, employ a scale-free type of architecture, they are relatively impervious to random damage but exceedingly vulnerable to the isolated failure of key components. In other words, most living neurons are expendable, but the rare few are not.

In brain networks, input and output neurons in particular are *not* expendable. Although they may also participate in the cognitive processes of the brain, input/output neurons (which make up only about a tenth of 1 percent of all human brains' nerve cells) come equipped with additional irreplaceable hardware not available to other nerve cells. The network's landscape memory can buffer it against the loss of hidden units, but not against the loss of specialized retinal neurons (input) or the motor nerve cells (output). Think back to the student network in which ten desks were equipped with photoelectric cells. We could remove any ten "hidden" desks from that network without hurting its performance, but if we removed the ten desks with photoelectric cells, the network became blind. The loss of even a few I/O neurons can cause catastrophic disruption of the network's ability to interface with the environment, causing (in the case of the brain) blindness, deafness, speech impairment, and paralysis. Most clinically devastating strokes injure these critical I/O neurons. Thanks to the brain's scale-free organization, strokes producing major long-term memory or cognitive deficits are relatively rare, since such deficits require the loss of massive numbers of hidden neurons.

This book assumes that *all* living ensembles operate within the network paradigm of computation; that's what makes them alive in the first place. This assumption can be challenged, at least for those living ensembles that we've examined thus far (bacterial communities, immune systems, the cytoplasm). Although I have no doubt that future research will confirm my belief that network architecture underlies all living systems, this belief can't be proven to everyone's satisfaction with the available evidence. However, the network architecture of nervous systems no longer seems open to debate. Network theory is based on neural behavior, after all, and an overwhelming amount of evidence supports the network design of animal nervous systems ranging from the primitive ganglions of worms to the giant cerebrum of Homo sapiens. The story of Alice and others like her is one form of evidence, but it's relatively weak. For stronger evidence, we need to explore the workings of the brain in more detail.

Let's begin with that ubiquitous structure, the neuron (this time, I finally mean a real nerve cell). Like the McCulloch-Pitts neuron, the typical animal nerve cell is a one-way I/O device. It receives many inputs from other nerve cells, adds them, then generates a binary on/off output when the sum of the inputs crosses the threshold for neuronal firing. The nerve cell differs from the McCulloch-Pitts neuron in that it doesn't stay "on" indefinitely. Instead, when the sum of inputs crosses a certain threshold, the nerve cell fires an electrical wave along its entire length. This wave, called an *action potential,* measures a little more than one-tenth of a volt and lasts for only a brief period, about a millisecond. After the action potential ceases, the nerve cell returns to its quiescent state and quickly recharges its batteries to prepare for the next action potential. If high levels of input still flow into the neuron, it will fire again after a short refractory period. A quiet nerve cell will have a low or zero firing rate, releasing rare action potentials; conversely, an active neuron will have a fast firing rate, releasing action potentials repeatedly in staccato fashion. We can define the nerve cell's output in two ways: short-term (whether or not it's firing an action potential at a given moment) or long-term (the firing rate, the number of action potentials released per unit of time). Generally, the firing rate is the more important of the two. As will be seen in the last chapter, brain networks convey information using both the magnitude of the firing rate (slow or fast) and its pattern (regular or chaotic).

Action potentials are generated by transient changes in the cell membrane's permeability to positively charged sodium and potassium ions. Simply put, the action potential is a burst of electricity—a chemical spark—that flows along an electrically charged membrane; the amount of charge carried by the membrane depends on the ability of the charged ions to enter or leave the cell. During an action potential, tiny ion "gates" open and close, thereby regulating the flow of ions through the membrane and altering its net charge. Opening and closing these gates takes time, a tiny delay that limits the rate at which neurons can fire. Furthermore, after firing an action potential, the nerve cell needs even more time to recharge. The chemical nature of the action potential makes nerve switches considerably slower than electronic switches like transistors and vacuum tubes. As previously discussed, the severe discrepancy between neuronal switching speed (fairly slow) and brain computational speed (incredibly fast) is strong evidence of the brain's parallel architecture.

Anatomically, the neuron has three parts: dendrites, cell body, and axon. The dendrites are treelike projections of the body that receive input from other neurons and pass them forward to the body. The body itself contains the cell's nucleus, cytoplasm, and other metabolic machinery. The axon, a single threadlike projection that can be several meters in length, relays the cell's output to other nerve cells or other target organs—predominantly muscle and glandular cells—at a rate of many meters per second. Action potentials flow in one direction, from the dendrites to the cell body down the length of the axon.

The axonal tip of every nerve cell reaches out to make contact either with other nerve cells or with some other target organ. The connection between an axon and another cell is called a *synapse,* from the Greek word *synapsis* (junction). There are two broad classes of synapse: electrical and chemical. Electrical synapses consist of a direct membrane-to-membrane contact between cells, resembling the soldered connections between wires on a circuit board. The transmission across electrical synapses is nearly instantaneous and flows in both directions, just as it would across a soldered joint.

When two nerve cells meet to form a chemical synapse, on the other hand, they don't actually touch. They come very close to touching—within a billionth of an inch—but no direct membrane-to-membrane contact takes place. If we look at the chemical synapse between an axon and a dendrite using electron microscopy, we'll see that a tiny gap, or synaptic cleft, remains between the axon's tip and the dendritic surface. On either side of this cleft, the cell membranes of the axon and dendrite have been specially modified to allow a highly localized form of chemical communication (synaptic transmission) to take place. Synaptic transmission, unlike transmission across electrical synapses, is a one-way street. Information flows from the transmitting neuron's axon (the pre-synaptic side) into the synapse, and then from the synapse to the receiving neuron (the post-synaptic side).

On the axonal (pre-synaptic) side of the synaptic cleft, the transmitting neuron's membrane contains minuscule packets, or vesicles, filled with a neurotransmitter. On the dendritic (post-synaptic) side, the receiving neuron's membrane is studded with various receptors contoured to bind the neurotransmitter. When the transmitting neuron fires, the action potential flies down the axon at high speed. When it reaches the vesicle-laden pre-synaptic membrane, the action potential causes some of the vesicles to burst, spilling their contents into the synapse. The neurotransmitter quickly (in nanoseconds) crosses the tiny cleft and binds to receptors on the receiving neuron's post-synaptic membrane. These receptors, once bound, electrically depolarize the membrane of the receiving neuron by altering the ion permeability of its membrane temporarily. If a sufficient number of synapses fire on the receiving neuron's membrane within a certain period of time, the cell's input threshold will be crossed and it will unleash its own action potential. A nerve cell adds synaptic input, just like the McCulloch-Pitts idealized neuron, by integrating the hundreds, perhaps thousands, of synaptic bursts it receives each second. If that sum depolarizes the membrane sufficiently, it will fire; otherwise, it stays silent.

Many chemical synapses also contain enzymes that quickly digest neurotransmitter molecules once the molecules enter the cleft. After the transmitting cell's action potential fades, these enzymes clear the synapse of any remaining neurotransmitter; the vesicles then quickly re-form on the axonal side and the synapse recharges to await the next action potential. The rapid degradation of the released neurotransmitter prevents a single action potential from having a prolonged effect on the receiving cell. Rather than digest them, some synapses recycle the neurotransmitter molecules, reabsorbing and repackaging them for later use. Both methods—digestion and recycling—severely limit the duration of synaptic transmission.

Think back to our party network, the corporate one filled with human guests. People could walk all about the room, but information exchange occurred only at close range via a transient form of communication called human speech, which dissipates quickly after it has been "released." Although neurons don't move, their communication still resembles party speech. They have to "talk" at close range across a narrow synaptic cleft using a language encoded in the neurotransmitter/receptor interaction. Like real speech, synaptic messages are temporally fleeting—the rapid digestion and recycling of neurotransmitters within the cleft guarantees that.

Chemical synaptic communication depends on a close molecular fit between transmitter and receptor, the same "lock-and-key" type of molecular communication used by enzymatic and immune networks. Neuronal "speech" is nothing more than protein biochemistry tidily packaged into countless synaptic reaction vessels. The brain is like a giant cytoplasm made up of interconnected chemical reactions, linked together by axons instead of by simple passive diffusion. In a single cell, reactions interact with each other directly, on a molecular scale. In a structure as large as the brain, however, passive diffusion and mixing alone won't work, because these physical processes are much too slow to drive an intelligent device weighing several pounds. Because mobile party neurons must "diffuse" through the crowd in order to speak to each other directly, coherent party networks can grow only to a certain size. Although a party of fifty mingling people may behave as a unified group, a party of ten thousand won't, at least not without some quicker way of transmitting information between the guests. In a party of ten thousand people (the average intimate Hollywood wedding, for example), coordinated action can't come simply from people milling about on their legs (too slow). Coordinated action among large-scale human networks requires the addition of cell phones or some other means of instant telecommunication. In the case of living systems, a monstrous cell like the freshwater amoeba may represent the upper limit of size a diffusion-driven enzyme network can attain before resorting to some form of hardwired communication. (Very large cells may even employ a kind of internal wiring consisting of protein microtubules.) By using axons and dendrites as phone lines linking small enzyme re-action chambers (synapses), enzymatic networks could soon grow to centimeter sizes and beyond, while still retaining the molecular quickness found inside a single cell. Nerve cells are simply large-scale surrogates for molecular diffusion, allowing enzymatic reac-tions to be linked quickly over centimeter scales.

Not all chemical synapses are the same. For example, different synapses can use different neurotransmitters. Generally, any one neuron emits only a single type of neurotransmitter (although there are exceptions to any general rule in the nervous system), but it may respond to multiple neurotransmitters emitted by other neurons. The nervous system uses many different neurotransmit-ters in its ten billion cells and trillions of synapses. If we think of the neurotransmitter/receptor complex as neuronal speech, each neuron *speaks* only one language, although it may *understand* sev-eral. The whole brain, on the other hand, is a chemical Tower of Babel.

A neurotransmitter typically consists of a short peptide (a chain of two or three amino acids) or some other small organic molecule, like an amino acid. Examples include glutamate, epi-nephrine, norepinephrine, acetylcholine, dopamine, and serotonin. The various neurotransmitters aren't distributed randomly in the brain, but rather, specific neurological functions are associated with specific neurotransmitters. For example, the neurotransmitter dopamine is used in certain motor control circuits located in the upper brain stem. Neurons that speak dopamine to other neurons are called dopaminergic; selective loss of brain-stem dopaminergic neurons, due to degenerative disease, viral illness, or the toxic effects of certain drugs, leads to rigidity and tremor of limb movements, a condition known as Parkinson's syndrome. The symptoms of Parkinson's can be partially relieved by administering L-dopa, a precursor of dopamine; the drug forces any surviving dopaminergic neurons to speak louder in order to compensate for their dead or mute colleagues. Norepinephrine and serotonin are among the languages of choice for neurons involved in mood regulation, and many antidepressant drugs target these neurons; acetylcholine is spoken in many memory and motor circuits, making it a favorite among Alzheimer's researchers.

To further complicate things, the receptors on the receiving (post-synaptic) side of a synapse also differ, even among synapses that use the same neurotransmitter. For instance, there are two broad classes of dopamine receptor in the human brain, and each has a different affinity for dopamine and dopaminelike drugs. The efficacy of synaptic transmission between a dopaminergic neuron and its target will depend, at least in part, on the type of dopamine receptor the target cell uses to detect incoming dopamine speech.

Dopamine certainly isn't the only neurotransmitter to have multiple receptor subtypes, but let's stay with the dopamine system for illustrative purposes. The first type of dopamine receptor, imaginatively labeled D1, has a lower affinity for dopamine than the second type, D2. Suppose a dopaminergic neuron fires the same action potential to two different synapses on two different target neurons, one synapse equipped with D1 receptors on the post-synaptic side, the other with D2 receptors. Suppose also that an equal amount of dopamine enters both synapses. The D2-equipped synapse transmits a louder message to the target cell, because its receptors bind dopamine more strongly. Remember: a neurotransmitter exists in the cleft only for a very brief period of time, and high-affinity receptors, which bind their targets more quickly, will bind more of the transmitter before it is cleared from the cleft. The use of receptors with varying affinities for the same neurotransmitter is one way the brain differentially weights its connections. (Different receptors for the same neurotransmitter may not even influence the receiving cell in the same manner—one class of receptor may open an ion gate, for example, while another activates a cytoplasmic enzyme. It all gets very complicated.)

In general, *the brain's connection weights are a function of synaptic efficiency* (efficient synapses mean strong connections) and synaptic efficiency depends, at least in part, on a neurotransmitter's affinity for its various post-synaptic receptors. The anatomic distribution of neurotransmitters and their receptor subtypes is largely hard-wired into the brain's anatomy at birth; this distribution pattern represents a property of the species and varies little among individuals. For instance, we all have dopaminergic neurons in the midbrain's substantia nigra, which project axons to target nerve cells in the globus pallidus equipped with D2 receptors. (None of this makes any sense to the average reader, so take my word for it.) The dopamine circuitry has been crafted at the level of the human genome over many millions of years, not learned in any individual's lifetime. And, because the nature of the receptors partially determines the brain's connection weights, the fact that we inherit a certain receptor pattern means that we also inherit certain "learned" information (learned, that is, by the species during evolution, not by any individual during his or her lifetime).

The electrical synapses have a fixed efficiency that is also genetically hardwired into the nervous system at birth. These synapses can't be modified during our lifetimes and don't participate in individualized learning at all. The brain's congenital pattern of neurotransmitter receptors and electrical synapses carries neural memories written in our very genes. We're born knowing certain things, but what things? What information is genetically programmed into our brain's connection weights from birth?

The obvious answer: instinctive behavior. Genetically programmed connection weights govern behaviors common to all individuals in the species. For example, electrical synapses abound in motor pathways controlling stereotypical actions like coordinated eye movements. A baby enters the world able to suck a nipple, a complex muscular response imprinted into our genetic connection weights. Although a baby must suckle immediately to survive, instinctive behaviors go far beyond simple eye movements and primitive survival reflexes. Genetically determined connection weights also provide the starting point for all future learning, *equipping the network with the general framework needed for acquiring more complex skills and knowledge later.* For example, I wasn't born knowing how to swing a golf club with precision, but I am genetically programmed to rotate my upper body in a swinging motion. I don't have to memorize the complex sequence of muscular contractions and relaxations that produce trunk rotation; that's inborn.

Every movement involves several muscles acting in concert. Merely raising my arm above my head requires the simultaneous contraction of the deltoid muscle, the pectoralis muscles, and the multiple muscles making up the shoulder's rotator cuff. The triceps muscle in the upper arm must also contract to prevent my hand from flopping down and hitting me in the head and, at the same time, opposing muscles that pull the arm back down must relax. When I lift my arm, I don't think about contracting and relaxing individual muscles; I think only of the arm motion as a whole. The complex patterns of arm movement form a "motor module" that has been pre-wired into my brain. The act of tossing a ball is largely hardwired into the brain; the more precise act of throwing a curve ball across home plate, on the other hand, still has to be learned. Nevertheless, it's much easier to learn how to throw a curve ball when you already know the basic throwing motion. Hardwired motor modules are the foundation, the starting point, for future learned behavior. In fact, all learning, even cognitive learning, may require some genetically determined set of connection weights as a starting point. Chimps, for example, may never be able to learn speech, because they lack the necessary speech modules in their brains. Likewise, we may never be able to fly, even with power-assisted wings, because we lack the motor modules for flight. The learning capacity of any brain, even one as complex and versatile as ours, is limited by its genetic design.

Thus, genetically imprinted connection weights give us a head start when we're learning new things. Many of the things we think we learn, like walking, have already been largely imprinted on the brain's connection weights at birth. The pre-wired connection weights are like rough drafts of a manuscript; evolution provides the words and we edit them as we learn. To summarize, the brain manifests two separate forms of weight matrix knowledge: (1) *instinctive* (genetic) knowledge determined by the genome and incorporated in the hardwired weights of electrical synapses and in the anatomic distribution of neurotransmitter receptors, and (2) *learned* (non-genetic) knowledge acquired during our lifetimes through the post-genetic modification of chemical synaptic weights. Unlike the immune system, the brain does not begin life as a complete tabula rasa. The genome provides a rough schematic for initial brain weights, detailing where different neurons will be located, what receptors and neurotransmitters those neurons will use, which neurons will be connected to the motor output and which to the endocrine output, and so on. The inborn connection weight matrix provides raw material for later adaptive learning. Non-genetic knowledge—that knowledge each individual accumulates in his or her lifetime—builds upon genetic knowledge by fine tuning genetically initialized connection weights, *sculpting inherited weight patterns into highly skilled, individualized actions and knowledge.* An instinctive pattern of arm movement can, after massive training, become a pitching motion worthy of a Cy Young award. (Or, in my case, it may become the incredible ability to toss popcorn directly into my open mouth—to each his own.)

Both genetic and non-genetic memory must be *trainable.* The training of instinctive brain weights takes place via mutation at the genome level—mainstream biologists call this evolution, but it's still a learning process that shapes the connection weights in response to external training. Non-genetic learning represents the moment-to-moment alteration of neuronal connectivity that allows us to remember what happened ten seconds ago, but I stress that geological evolution is a form of long-term memory, too. The human genome recalls what happened to our lineage several million years ago. My brain contains both the connection weight changes that arise from personal experience and the changes acquired by the aggregate experience of Homo sapiens.

Weight training at the genome level—evolution—represents genetic network learning, while the direct changes in network weights that occur during a network's lifetime represent non-genetic learning. As noted previously, the first living networks could use only non-genetic learning (because there weren't any genes). After the advent of the genetic code, networks began to rely almost exclusively on evolutionary learning. Although simpler organisms have some capacity for non-genetic learning too, that capacity is limited; nevertheless, it exists—slime molds, for example, can learn to navigate a maze in search of food. That said, most learning at the microbial level, as was seen in chapter 1, must occur at the gene level. This doesn't make microbes stupid, since they've managed to accelerate their genetic learning greatly.

As will be discussed shortly, *chemical synapses drive non-genetic learning in the brain.* Unlike electrical synapses, chemical synapses can be readily modified in response to external training. In fact, the origins of non-genetic brain learning can be traced to the ancient evolution of the chemical synapse. Microscopic examination of the simplest invertebrate brains reveals that even these pipsqueak organs use both electrical and chemical synapses. Creatures as "simple" as ants have a capacity for learning. As Edward O. Wilson points out in his excellent book *The Insect Societies,* separating learned from instinctive behavior can be difficult in creatures with limited behavioral repertoires; nevertheless, American psychologist Theodore Schneirla showed that Formica ants can, like slime molds, be taught to run mazes. In fact, ants are almost as good as laboratory rats in this regard. Honey bees memorize the location of food sources and convey that information to their colleagues using ritualized dance movements. Still, most simple invertebrate nervous systems are predominantly instinctive networks that rely more on reflex and less on individualized, learned behaviors. Simpler brains illustrate how post-genetic learning must be layered on pre-existing, genetically determined connection weights. Ants and bees can be taught individualized acts, but those acts never stray far from in-born ritualistic behaviors. The ant can be taught a maze because running a maze resembles foraging for food.

Insect brains resemble the first perceptrons in that their weights can be trained, but their networks lack significant numbers of hidden neurons and so have a limited capacity for associative learning. Although even the most rudimentary nervous systems, like that of the gastropod mollusk Aplysia, can manifest a crude version of associative learning, only vertebrate brains harbor large numbers of hidden neurons. Hidden neurons allow brains to store information that appears unrelated to immediate survival. We can write symphonies because of our hidden neurons; honeybees can't. Unfortunately, our decreased dependence on instinctive behaviors also means we're not as proficient at them. Vertebrates must reinforce their instinctive connection weights at a young age by engaging in mock survival activities. In mammals we call this "play," and, although entomologists have searched for it for years, no evidence of play has yet been documented in insects, even in social insects.

We can roughly divide nervous systems into two classes: (1) invertebrate brains, which have few hidden neurons, genetically determined connection weights, and a limited capacity for individualized learning, and (2) vertebrate brains, which have many hidden neurons, highly trainable connection weights, and a great capacity for individualized learning. Invertebrates are so good at instinctive behavior that they're born knowing nearly everything they need to know, while vertebrates may be so poor at instinctive behavior that they have to learn and practice the most basic survival skills during their lifetimes. Once again, we should avoid making value judgments about this particular Yang/Um split in brain design. One form of brain isn't "better" than another—it's just right for a given animal.

Like bacteria and immune systems, insects and other invertebrates exploit their accelerated evolutionary pace in order to adapt their neuronal connections quickly. Their relatively short life spans and lack of hidden neurons limit their potential for non-genetic learning. In animals with a very slow evolutionary pace, like us, genetic learning can't do it all. The NFL doesn't want to wait for a good quarterback to evolve; they need someone with superb throwing skills right now. Thus, the evolution of big brains compensates for the slow evolutionary pace of vertebrates.

We can understand the depth of our genetic knowledge by examining diseases that rob us of it. Parkinson's disease, for example, strips us of our genetic knowledge of motor control. By disrupting inborn motor regulatory connections, the illness deprives sufferers of more than four million years of bipedal learning. That's what makes such illnesses difficult to treat. If a man forgets a learned task, he simply has to relearn it. But when a Parkinson's patient forgets how to move at the instinctive level, those motor instincts can't be replaced in one lifetime. Rehabilitation centers can teach new skills, but they can't replace the process of evolution; they can't retrain the genome. The Parkinson's sufferer's brain forgets the basic patterns of movements. The motor modules fall apart and the brain becomes confused, unable to coordinate the simplest tasks. The victim can't recall which muscles to contract and which to relax, and in what sequence, causing movements to become stiff and jerky as opposing muscles fight to control the same limb. In the end stage, the brain can't initiate movements at all and the victim descends into inertia.

Conversely, Alzheimer's disease dissolves the *learned* connection weights, disrupting the chemical synapses that have been post-genetically modified to encode things like the names of our childhood pets. As the disease progresses, the victim is left with only the remnants of the inborn, instinctive connection weights. Primitive reflexes present only in infants and overridden in early childhood, when post-genetic learning takes over, begin to reappear. Stroke the cheek of an end-stage Alzheimer's patient and, at age ninety, he may turn his head reflexively in search of his mother's nipple.

Genetic learning requires the alteration of connection weights through modification of the genes encoding those weights. Evolution also provides other ways for the heritable alteration of network behavior, such as increasing the number of neurons (brain size) over time. The genome encodes all of the brain's hardware, including the baseline connection weights. The evolutionary progression from small brains to large brains, like the progression from non-adaptive immunity to adaptive immunity, sprang from the need for a greater degree of non-genetic learning capacity. Genetic and non-genetic learning capacity appear to be inversely related: the simplest organisms (viruses) have virtually no capacity for individualized learning yet can evolve genetically at light speed. Large organisms, like vertebrates, possess massive reservoirs for individualized learning, but their immense complexity makes them ponderously slow in the genetic realm. Bacteria reproduce in minutes; humans in decades. Fruit flies lay thousands of eggs in a day; a human female produces a few hundred eggs in her lifetime. Different species choose to partition their intelligence in different ways, some adopting a predominantly genetic approach, others going the non-genetic route. We can speculate that, in any given ecosystem, biological intelligence is a zero-sum game: the more genetic intelligence a species possesses, the less non-genetic intelligence (and vice versa).

I've gone out on a limb many times in this book, and so I'll crawl out even further and formulate a speculative law: *the sum of a species' genetic and non-genetic learning capacities is the same for all species on earth.* Bacteria have scant non-genetic learning capacity but a huge genetic learning capacity; in humans, the reverse is true. We insist on having one child at a time, on average, and with such a low reproductive rate (and a generational time measured in decades, not hours or weeks), we need very big brains indeed. The supremacy of human intelligence is reduced to a mirage, an artifact of the parsing of network learning between genetic and non-genetic learning skills. All life on earth may be smarter than it was a billion years ago, but all species alive today are co-valedictorians of earth's survival school, class of this year. Theologically, I can make no more inflammatory statement, but there it is.

(Of course, proving the above conjecture is currently impossible, since we have no way of rigorously measuring the intellectual capacity of individual organisms, let alone whole species. Theoretically, it may one day be possible to compare the intellectual capacities of different organic networks in a more quantitative fashion.)

Nature has no need for excess intelligence. We know from computational theory that intelligence doesn't come free (or even cheaply), either in living networks or in human-made machines, and so selection pressures will work to keep total intelligence to the minimum required for survival needs. A species should have the intellectual capacity needed to solve the problems it faces and not much more. And, since all species face basically the same problem set—survival—they all have to be comparably intelligent, give or take a few neurons here or there. Since most biologists tend to think of evolution as stupid, they routinely neglect to include genetic intelligence in the overall equation, making non-genetic learners like us look like comparative geniuses. Well, we *are* geniuses, but so are all other species alive today. This book is not a condemnation of humankind, but a celebration of life's brilliance.

How does non-genetic learning take place in the brain? To understand this process, let's examine the most common type of chemical synapse in the human brain, the *glutamatergic* synapse. The glutamatergic synapse uses the neurotransmitter glutamate; more than half of all brain synapses are of this type. At the receiving end of the glutamatergic synapse, the post-synaptic membrane possesses two subtypes of glutamate receptor: the N-methy-D-aspartate (NMDA) receptor and the Q/K receptor. (The origin of these tongue-twisting names is of no importance here.) Both receptors bind glutamate, but they work in very different ways. The Q/K receptor is rather mundane, so we can promptly forget about it. The NMDA receptor, on the other hand, has a special quality: *it works only when both the transmitting neuron and the receiving neuron are firing action potentials at exactly the same time.* The cells' simultaneous firing may be purely accidental, but that doesn't matter; the NMDA receptor works only when both cells are active at once.

In electronic parlance, the NMDA receptor acts as a coincidence detector. So what's the big deal about that? As it so happens, an activated NMDA receptor does more than just deliver a transient synaptic message at freakish intervals. *Every time it becomes active, the NMDA receptor increases the long-term efficiency of the synapse by a tiny bit.* By enhancing the synaptic efficiency, the NMDA receptor strengthens the connection between two neurons that happen to be simultaneously "on." This mechanism, discovered in the early 1990s, is now called *long-term potentiation,* or LTP, but we already know another name for it: Hebbian learning—the tendency of active neurons to strengthen their connections with other active neurons. Forty years after he first proposed his learning rule, researchers finally proved Hebb was correct. We now understand how his rule works on a molecular level, at least in brains. (Interestingly enough, the glutamatergic synapse may become less Hebbian as we grow older due to age-related chemical changes in the NMDA receptor. This may be one reason why learning is easier and more rapid in children.)

Hebb's rule, in its original form, has one drawback: it provides for the *strengthening* of connections between active neurons but doesn't explicitly state how connections can be *weakened* with disuse. So far, I've assumed that Hebbian learning also includes some way for inactive neurons to weaken their connections. Otherwise, if the original Hebbian rule alone dictated a network's behavior, it would eventually become so saturated with strong connections that no learning could occur. This limitation of Hebb's rule was recognized long ago, and most modern applications of the rule assume that inactive connections tend to weaken over time.

In the introduction to this book, I asserted that all intelligence is intrinsically Darwinian. Darwinism assumes that some form of competition takes place, and there is indeed competition among nerve cells in the brain. In the cerebral cortex, neurons are arrayed in discrete layers. Within each layer, cells are connected to their closest neighbors using NMDA synapses and are connected to more distant cells via inhibitory connections that use a neurotransmitter called gamma-amino-butyric acid, or GABA. GABA is a unique neurotransmitter, in that it doesn't stimulate the receiving nerve cell to release an action potential (like glutamate and other neurotransmitters do), but actually *dissuades* the cell from firing.

Remember, before firing an action potential, the neuron must add all of its inputs to determine whether that sum crosses the threshold for firing. GABA synapses provide *negative* input to the cell; a neuron must *add* all incoming NMDA signals and then *subtract* the GABA signals to calculate total input. When a given neuron fires an action potential, it sends positive NMDA signals to its closest neighbors and negative GABA signals to its more distant neighbors.

When information begins flowing into a cell layer, neurons begin firing away, groping for a piece of the action like stockbrokers screaming on a trading floor. Trading alliances soon form among neighboring traders, and those alliances then seek to squelch all competition elsewhere on the floor. Perhaps a sexual analogy works better (it usually does). If we think of an action potential as a desirable thing—a neuronal orgasm—a firing neuron rewards its closest comrades with stimulating NMDA foreplay while sending cold GABA showers to its distant colleagues. Because the synapses are all subject to Hebbian training, a pattern of rewards and inhibitions soon becomes ingrained in the cortex. Friendly cells will strengthen their own connections and weaken the connections among their competitors elsewhere in the layer. When a new pattern of inputs arrives at the cortex, neurons begin fighting for that information all over again. Eventually, the cortex settles into patterns, where tiny islands of winning neurons enjoy orgies while other islands of loser neurons sulk alone on their couches, watching videos and eating cheese curls. These patterns contain the information we know as long-term memory.

When I type words on a page, I create islands of black ink set against a background of white. For an image to be visible, there must be some form of contrast between light and dark, black and white (Yang and Um revisited). The NMDA/GABA dichotomy in the cortex provides the necessary contrast between winning and losing neurons, a contrast that then can be used to store patterned images within neuronal layers. Repeated exposures to patterned information will sharpen the contrast between neuronal winners and losers, effectively "focusing" the image. Nearly twenty years ago, Finnish engineer Teuvo Kohonen created a neural network model of cortical behavior based on a competitive architecture. Now known as the Kohonen network, it can form self-organizing feature maps that extract and store key features from a set of patterned data. Like other human-made network models, the Kohonen network can't approach the intricacies of the human neocortex, but it has provided some insight into how neuronal layers handle patterned data. For example, Kohonen networks detect recurring features in different data sets. If presented with a completely new set of data each and every time—data with no common features—the network can't learn. When presented with data sets with some shared features, the network can recognize and extract those features. Like a Kohonen network, my brain can look at a group of barnyard animals and extract common features: four legs, a head, a tail, and so forth. Kohonen's model is far too simple to be taken as a serious model of the human cortex, but it does give us some insights into how the cortex learns.

We come back to a common theme: to produce biological intelligence, create an ecosystem and force its inhabitants to compete for some limited resource. In the student network, the ecosystem was a set of desks and the resource was electric current. In the immune system, the ecosystem was an immune system and the resource was an antigen. In the human brain, the ecosystem is the cortex and the resource is, once again, electrical energy. Like the protagonist of the science-fiction story I described in chapter 1, Nature sows competition in a restricted ecosystem and then harvests the intelligence it produces like so much wheat.

So far we've assumed that the brain utilizes some form of attractor-based memory. Attractor-based memory is equivalent to the content-addressable memory discussed in an earlier chapter. Unlike serial computers, which simply stuff chunks of data into numbered addresses inside a memory register, network devices store information in attractors, and attractors, by their very nature, store related memories—memories that share common attributes.

The "address" in an attractor memory is partially determined by the *content* of the memories themselves. Is there any evidence that the brain uses content-addressable memory? Yes. In a series of experiments, Yasushi Miyashita taught a set of unrelated visual patterns to monkeys by presenting the patterns in a certain order. After the training was complete, the monkeys were again shown the visual patterns, this time in a random order. As the monkeys saw the patterns again, Miyashita measured the electrical activity in cortical nerve cells using miniature electrodes. He discovered that the spatial patterns of neuronal firing corresponded to the orderly sequence in which the visual patterns were first learned. In other words, two patterns that were close together in the learning sequence were stored in a similar location in the brain. The primate cortex lumped several patterns into one attractor using the only "content" they shared, namely the fact that they were learned at about the same time. A woman who smells movie popcorn and remembers her first date over thirty years earlier is manifesting a similar form of content recall.

David Freedman and his colleagues at the Massachusetts Institute of Technology provided a much more recent demonstration of content-addressable memory in primate brains. Freedman was interested in perceptual categorization, the process of assigning perceived objects into rigid groups regardless of their physical appearance. Humans place physically dissimilar apples and bananas in the same category (fruit), but exclude red billiard balls from that category even though they resemble apples. Freedman trained monkeys to recognize computer-generated images of cats and dogs and then recorded the neuronal firing rate of prefrontal cortical cells as the trained animals were again shown digital animal images. As the MIT group reported in early 2001, monkey neurons aligned according to the *category* of the image (either catlike or doglike). A "cat" neuron responded more strongly to catlike images, and "dog" neurons responded in kind to doglike images. In a previous chapter, I speculated that there was a "Beatles" area in my brain's landscape. Freedman and his coworkers showed that data categories can indeed be mapped onto specific locations of the cortical landscape.

Thanks to Hebbian learning and the deployment of GABA-ergic inhibition, neuron is pitted against neuron in a Darwinian competition for information. In the immune system, the competition was for antigens; in the brain, the competition is for action potentials. Lymphocytes form alliances and strike down competitors in order to maximize their share of antigens, and neurons do likewise to maximize their share of incoming electrical signals. In both instances, the competition leads to optimal behavior of the communal system taken as a whole.

But wait. Don't these "competitions" seem somewhat contrived? Nerve cells don't actually "die" (well, sometimes they do—neuronal death during normal competition for nerve impulses has been linked to dementia, including Alzheimer's, but that's a speculative topic beyond our present discussion). How does the faux competition among nerve cells and lymphocytes compare to the real "life and death" struggle that plays out in the ecological world? Consider the case of tournament fencing. Fencers use the same types of weapons, and the same techniques, that they would use in a real fight to the death. Of course, they don't fight to the death, but when a victor thrusts his or her weapon into the shielded breast of an opponent, that opponent would, under normal circumstances, be killed. Thus, for all intents and purposes, a fencing match isn't any different from a real sword fight, except that the equipment and rules have been slightly modified so that the same fighters might be reused again and again.

Likewise, organ systems like the brain exploit the computational power inherent in Darwinian selection, but they modify the equipment and the rules so that the same players may compete again and again. Theoretically, however, there is no difference between the "evolution" that takes place in the immune system after a vaccination, or in the brain during rote memorization, and the evolution that turned single cells into human beings.

For true evolution to occur in the competitive ecosystem of the brain, however, some random element must be introduced. As we already know, genetic evolution depends upon random mutations and other errors in DNA information handling. Another name for random errors in data transmission is *noise*. We tend to think of noise as bad, a thing to be cleansed from our expensive car stereo systems, but in Darwinian networks, noise becomes indispensable. There could be no evolution without some level of nucleic acid noise, for example.

Noise provides a Darwinian network with the raw materials for creativity. Earlier, I compared the random agitation of activation states and connection weights to Brownian motion. This Brownian motion allows the network to modify its weights even when it isn't being trained. The noiseless student Hopfield network modifies itself only in response to external training, but a noisy network modifies itself continuously. The student network can learn and associate patterns of input with other patterns of output, but it can't think. After the teacher has gone home, it can't figure out new ways to correlate inputs and outputs all by itself. *The process we call thought results from the Darwinian sculpting of network noise in the absence of external training.*

Theoretically, the ratio of signal-to-noise is critical to network performance. Here, "signal" means the true value of the data in question. If a connection weight or neuronal activation state has a true value of 1.0 and then randomly fluctuates over time by plus or minus 0.01, its signal-to-noise ratio will be 100. A network possessing a very high signal-to-noise ratio—little random error—can't think very well; it's too "concrete." Conversely, a network with a low signal-to-noise ratio—a large amount of randomness—becomes too abstract; it has difficulty making consistent associations between input and output. For example, when the random errors in the connection weights begin to approach the values of the weights themselves—a signal-to-noise ratio approaching 1.0—the network's "thoughts" will overshadow any weight changes induced by environmental training. The network will become preoccupied with internal imagery, and reality will have no meaning as the network disconnects itself from the outside world. In other words, it will become psychotic. This is a difficult concept, so a few illustrations are in order.

Remember the little pinball game, the one with the tiny balls rolling into holes? The level of "noise" in that game is determined by how hard we shake it. If we try to steer the ball precisely, we are solving the game with very little random noise (a high signal-to-noise ratio). As we agitate the game, we raise the amount of noise and lower the signal-to-noise ratio. The key to solving the game lies in determining just how much noise to inject into the system. If we don't shake it at all, the game proves too difficult. If we shake it too hard, however, the noise (shaking) overwhelms the signal (our attempts to steer the balls) and the balls never come to rest in any of the correct holes. Different problems require different levels of noise; an intelligent network must be able to regulate its signal-to-noise ratio. It must know how hard to "shake" itself in order to arrive at the correct solution.

Consider another example from the world of advertising. Years ago, a maker of chocolate and peanut butter candies ran a

television commercial in which two people, one carrying a chocolate bar and another a jar of peanut butter, collided. As they
picked themselves up, they noticed that the chocolate and peanut
butter had become jumbled together. They were angry at first,
until they tasted the result. In this case, noise (the accidental collision of two people) yielded a new and useful association (chocolate and peanut butter taste good together). In a perfect noise- and
error-free world, no accidental associations can occur, good or
bad. The essence of associative thinking is to mix things up and
then separate the good from the bad (I doubt that the peanut butter lover would have been as pleased had he run into a man carrying sardines). The key is to modulate the degree of noise and
error. Too little noise means no associations at all; too much noise
yields nothing but nonsense. The act of thinking injects noise
into the system, allowing old patterns to "collide," forming new
associations.

Neurons and their synapses are noisy, orders of magnitude
noisier than their electronic, human-made counterparts inside
computers. This fact alone suggests that brain noise plays some
useful role; otherwise, genetic selection pressures would have
eliminated it (or at least reduced it). However, the signal-to-noise
ratio, being so very critical to the balance between concrete thinking and imaginative thinking, must be regulated. In the human
brain, the signal-to-noise ratio comes under the control of neuro-
modulating systems that use, among other things, our old friend
dopamine. These dopaminergic pathways permeate the brain and
control the ambient signal-to-noise ratio.

One way of clinically measuring the signal-to-noise ratio in
humans is *semantic priming*, a psychological testing tool invented in
the 1970s. The test exposes subjects to pairs of words on a computer screen. The first word is flashed for a fraction of a second,

just long enough for the subject to read it; then a second word appears briefly on the screen, and the subject is asked to determine quickly if that second word is really a word or simply a nonsense string of syllables. The mean reaction time for making a correct response is calculated after a set of word pairs. Wrong responses don't matter; the subjects are encouraged to be right only so they will delay long enough to be certain of their answers.

In normal subjects, the reaction time becomes shorter if the two words are related in some way. For example, it will take me—presuming I'm normal, which is by no means certain—less time to recognize the word *down* when it is presented after *up* than it will to recognize *banana* when it is presented after *automobile*. The first word primes the pump, so to speak, for a related word. The mean difference between pairs of related and unrelated words is called the *semantic priming effect;* it normally ranges between 30 and 50 milliseconds. In a system free of noise, the semantic priming effect should be zero, because noise-free networks have a sharp, literal view of the world and can recognize all correct words equally quickly, even in the absence of priming. In a noisy system, the impact of external data is less concrete and the network must rely more on internal imagery to deduce correct answers. This slows down the process. Thus, the size of the semantic priming effect is a measure of the brain's signal-to-noise ratio, at least as it applies to language processing. The oral administration of L-dopa, a drug that increases global brain dopamine, reduces the semantic priming effect by increasing signal, suggesting that the brain's signal-to-noise ratio comes under neurotransmitter control.

In cases of immediate emergency—that dog coming to attack me—the signal-to-noise ratio increases, and thought becomes focused and non-abstract. Find exit, run away. There's no need for abstract association in this circumstance. Running into someone with a jar of peanut butter would serve no purpose here—I want only to survive, not to be creative. The most urgent responses (reflexes), like pulling my hand away from a hot stove, bypass all of the brain's associative neurons entirely. In less dire circumstances, on the other hand, the signal-to-noise ratio decreases and our thinking becomes more associative, allowing more creative solutions. Einsteinian geniuses may harbor the reversible ability to drive their signal-to-noise ratio very low, into the fantasy range, freeing their thoughts from the crippling bondage of external experience and allowing them to think more abstractly than the general population.

Many psychoactive drugs exert their mind-altering effects by increasing the influence of neuronal noise, thereby allowing internal imagery to overwhelm our sensory inputs. In the same vein, two researchers from my hometown of Pittsburgh, Jonathan Cohen and David Servan-Schreiber, proposed nearly a decade ago that schizophrenia, a delusional disorder affecting almost 1 percent of the general population, results from the failure of dopamine pathways to regulate the brain's signal-to-noise ratio. Not surprisingly, the semantic priming effect is larger in schizophrenics relative to the normal population, and antipsychotic drugs like thorazine have been designed to alter the brain's dopaminergic pathways.

Mutant mice born with vastly reduced numbers of functional NMDA receptors exhibit odd, repetitive behaviors that resemble those found in human schizophrenics. Without NMDA receptors to carry signal, the mouse brain's signal-to-noise ratio falls and the mouse becomes self-absorbed. In humans, the depletion of NMDA receptors can be accomplished pharmacologically, using the drug phencyclidine ("angel dust"), which temporarily disables glutamatergic synapses. An overdose of angel dust renders the user acutely psychotic—I've cared for several cases of phencyclidine abuse and can attest to this personally. Some authors now believe that schizophrenia may be caused by NMDA receptor malfunction. Like the dopamine theory, the NMDA theory points a finger at the signal-to-noise ratio in the psychotic (or intoxicated) brain.

There has always been a fine line between delusional thinking and creative genius. That line depends on the signal-to-noise ratio in Darwinian networks. Back in chapter 1, I referred to colonies of bacteria with high rates of mutation (hypermutators) as microbial Van Goghs—creative but a little insane. The mutation rate in microbes (indeed, in any genetically evolving network) determines their signal-to-noise ratio; hypermutators are schizophrenic—they have a low signal-to-noise ratio. Cancer cells may be viewed similarly; deprived of input neurons (cell receptors), their cytoplasmic networks must operate with a lower signal-to-noise ratio too.

Recall that the rate at which DNA is *damaged* by environmental toxins is fairly constant, but cells can *repair* that damage as quickly or as slowly as they want. The actual mutation rate equals the rate of DNA damage minus the rate of DNA repair. All cells possess enzymes capable of detecting and curing genetic injuries. By switching repair enzymes on or off, the effect of environmental mutations can be lowered or raised. Theoretically, a cell could reduce its mutational noise to zero—but it doesn't. All living networks need noise, and they retain it even when they have evolved the technology for eliminating it altogether. (As we'll explore in more detail later, noise prevents networks from becoming trapped in one stable state forever. Unlike serial computers, which move sequentially from one logical operation to another, networks must reach solutions by roaming through a landscape. Noise provides a kind of momentum, agitating the network toward an optimal stable state.) All networks also exert control over their signal-to-noise ratios; in the immune system, for example, the signal-to-noise ratio falls under the control of immune modulators like the interleukins.

Let's pause here and examine the meaning of the signal-to-noise ratio in more concrete terms. Consider two images of a French landscape, one created photographically using high-speed film on a sunny day, the other created by artist Georges Seurat, who painted dreamlike images using fine dots of color. The crisp photograph has relatively little "noise" and hence a high signal-to-noise ratio, leaving little to the imagination. The Seurat painting, on the other hand, is indistinct and noisy, with a low signal-to-noise ratio. If we place the two scenes side by side, we will have a graphic illustration of high versus low signal-to-noise ratios. Suppose now that we wish to find a small dog running through the landscape. The image of the dog constitutes information, a signal that we must extract from the image. In order for us to see the dog, the signal of its image must rise higher than the level of noise. In the photograph, the noise level is so low that even a small dog should be readily apparent above any background blurring. In the Seurat painting, however, the image quality may be too noisy— too blurred—to allow the signal image of a small dog to emerge. Of course, even the photograph has some fixed level of noise. For example, we couldn't see a bacterial cell in a landscape photo, no matter how much we enlarged it. The signal-to-noise ratio determines the amount of meaningful information we can extract from a given image or set of data.

It would seem desirable to eliminate noise altogether, but then we wouldn't have the beauty of Seurat paintings. Our galleries would be filled with only photographs. Likewise, in computational theory, as in art, we need noise to stimulate the imagination.

No creativity can emerge without some degree of noise and uncertainty.

Consequently, network dysfunction isn't merely a matter of having a too-high or too-low signal-to-noise ratio; it's more a matter of a ratio unsuited to situational needs. Children, because they need to acquire patterned associations more quickly than adults, tend to have a lower signal-to-noise ratio, making them more prone to imagination and less reliant on external reality. In other words, children tend to see the world as a Seurat painting, while adults tend to see it as a photograph. To paraphrase Samuel Beckett, we all start out life as schizophrenics; a small percentage of us stay that way into adulthood.

Maintaining a high signal-to-noise ratio requires a greater expenditure of energy. Remember Maxwell's demon: information is energy. Suppose I'm at a party with a lot of background noise— music, the clattering of glasses, the drone of distant conversations. A man approaches me and begins whispering. That person's speech becomes a signal that I must try to distinguish from the background noise. If he speaks very softly, the signal-to-noise ratio is too small for me to make out what he's trying to say. I must ask him to speak up, and he does, increasing the volume of his voice. This causes the signal (his speech) to rise above the level of the background noise so that I can now comprehend him. However, increasing the volume of his voice takes more energy. The human voice is no different from any other sound-generating device in that it needs more energy to generate a louder sound—a radio played at a loud volume will expend its batteries faster than a radio kept at a low volume. To increase the signal, we must use more energy.

In sleep states, the signal-to-noise ratio falls as the energy consumption of the brain decreases, allowing greater freedom of association. The bizarre visual images that occur during our dreams may be physiologically related to the "flight of ideas" and hallucinations of the schizophrenic's wakeful brain. Dreaming is an extremely shaken state. The ball inside our head bounces from one attractor basin to another, causing weird, unexplainable juxtapositions of the dream world (like the time I went to dinner with The Incredible Hulk and my old clarinet teacher).

But enough about noise for now—we'll revisit the subject in more depth later. Back to the brain. Is the brain (or, more specifically, the brain's cortex) a purely self-organizing network? Remember, there are two broad classes of networks: those that learn entirely by themselves (self-organizing networks) and those that require additional hardware to help train them (supervised networks). Self-organizing networks modify their connections internally (local learning). In supervised networks, on the other hand, connection weights are modified, at least in part, by some external training device. Recall that Robert's golfing robot used a form of nonlocal, or supervised, learning (backprop) that was implemented by a training device external to the network itself. Psychologist James McClelland and others have proposed that the hippocampus, a structure on the inner aspect of the temporal lobe, may serve as a kind of trainer, modifying the cortical connections during learning.

The word *hippocampus* is Greek for "seahorse"; the structure's name comes from its curlicue configuration on cross-section, which resembles the little animal's tail. The hippocampus plays a huge role in short-term memory, the recollection of events from minutes or hours ago. Bilateral destruction of the hippocampus produces a bizarre syndrome in which the victim can recall past events but can't retain the memory of any recent events at all. The most famous example of this disorder was the legendary patient

H. M., a man who underwent temporal lobectomies for seizures in the 1950s and subsequently became something of a sideshow attraction. A more recent (although fictional) example was the comic character "Mr. No-Short-Term-Memory," played by actor Tom Hanks on "Saturday Night Live" a few years ago. The audience guffawed as Hanks failed to recall even the bite of food he had just placed in his mouth, but the affliction is no joke. I once helped create a Mr. No-Short-Term-Memory myself. During the removal of a colloid cyst, a benign but life-threatening tumor that occurs deep in the brain, a colleague and I damaged both of the patient's fornices, thin cables that provide the outflow from the hippocampus. In the days after his surgery, a conversation with the man would go something like this:

Mr. G.:	Doctor, what time is it?
Me:	It's noon, Mr. G.
Mr. G.:	Lunchtime?
Me:	Yes.
Mr. G.:	What time is it?
Me:	Noon.
Mr. G.:	Time for lunch?
Me:	Yes.
Mr. G.:	I'm hungry, is it time for lunch?

Mr. G. would read the same sports section for hours, always returning to the front page as if he had never seen it before in his life. Thankfully, he eventually recovered. It's one thing to theorize about such conditions, but quite another to be responsible for them personally.

The hippocampus is a staging area, where short-term memories are processed and prepackaged for later storage as long-term memories in the cortex. Primate experiments demonstrate that short-term memories stay in the hippocampus for several weeks before finally being exported to the cortex. The hippocampus transfers memories to the cortex by training cortical neurons in some way, either by directly modifying their synaptic connection weights (like a backprop trainer) or by simply exposing the cortex to the same patterns repeatedly and allowing Hebbian learning to take place (like the teacher of our student network), or a mixture of both. The hippocampus remains a mysterious structure. It must be a Hebbian network itself, since hippocampal synapses are among the most "Hebbian" in the brain (the mechanisms of LTP—long-term potentiation, the chemical basis of Hebbian learning in the nervous system—were discovered in hippocampal slices). The precise mechanisms by which short-term memories become long-term memories remain unknown.

One thing becomes apparent as we dissect the network organization of the nervous system: it's not one network, but many networks interleaved to form a single device. Each of the brain's component networks likely possesses a unique architecture and learning behavior; how the brain's many networks work together remains unknown. In truth, we've barely begun to understand the magnitude of the brain's complexity. The leap from simple network models to the mammalian brain is immense and beyond our imaginings.

Our current network models are pale imitations of a living brain. Computer simulations of networks usually employ hundreds, perhaps thousands, of simple sigmoidal neurons, each receiving a handful of inputs. But contrast this with the brain, which uses ten billion highly sophisticated neurons, each receiving up to eighty thousand inputs. Furthermore, a mammalian nerve cell has the computational power of a modern laptop or more, a far cry from a transistor or simple switch. Finally, in our artificial networks, connection weights consist of variable resistors or software links trained by some relatively simple algorithm. In the brain, the weights consist of chemical synapses of enormous subtlety and complexity, trained by learning algorithms far more sophisticated than any now known.

In order to keep this chapter concise and so not be accused of my own brain chauvinism, I've had to simplify things enormously. For example, when I said that D1 and D2 receptors have different affinities for dopamine—well, that's true and not true. The affinity of neurotransmitter receptors varies according to a number of factors and isn't easy to define in absolute terms. The same receptor molecule may have different affinities at different times. A complete analysis of neurotransmitters and their receptors would fill three books this size.

In his recent monograph, *Circuit Complexity and Neural Networks,* Ian Parberry notes that we are just beginning to comprehend how simple networks "scale up" to become giant networks. The study of giant networks, including brains, lies within the scope of an emerging paradigm known as computational complexity theory, which, along with statistical thermodynamics, is one of those fields that requires an IQ above 170 to fathom completely. This field may someday explain how collections of nerve cells yield that thing called consciousness, but, for now, the intricacies of the brain can't be accurately modeled by our puny tools.

Nevertheless, the brain, like all living things, is still a network, a network that just happens to be "wider than the sky." (So I am a little biased—it's still my bread and butter, after all.)

8 SUPERORGANISMS

*"...the insect society...is so strikingly analogous
to the Metazoan body regarded as a colony, or indeed to
any living organism as a whole, that the same general
laws must be involved....We can only regard the
organismal character of the colony as a whole as an
expression of the fact that it is not equivalent to the sum
of the individuals but represents a different and at
present inexplicable 'emergent level'..."*

—William Morton Wheeler, *The Social Insects*, 1928

There is a creature that lurks in every backyard; in fact, the average backyard harbors a dozen of them or more. At birth, this beast may be only a few millimeters long, but by the age of sexual maturity—about five years—it will reach its final adult size of two to three meters. It will consume plants, seeds, and insects, or, amazingly enough, may even cultivate its own food. It lives most of its life underground, largely out of human sight, where it struggles through childhood, matures, reproduces, grows old, and finally dies a natural death like any other animal. These creatures have a typical life expectancy of about fifteen years and, although we usually have nothing to fear from them, they can turn into killers if provoked. They're hardly mysterious. On the contrary, they're among the commonest forms of animal life on this earth: ant colonies.

Take note that the description above doesn't apply to *ants*, but to ant *colonies*. The colony has a life expectancy of fifteen years, but the typical ant may live only a few weeks. The colony reproduces, not individual ants, and in nearly every biological respect (growth, behavior, reproduction), the ant "organism" can be considered to be the colony, not the ants themselves. Two different colonies of the same species will still be genetically different, just as two humans are genetically different, and will manifest different behaviors that can be influenced by natural selection. Thus, not only do the ants evolve, but their colonies evolve as well.

Colonies manifest a collective intelligence. If a nest becomes contaminated with mold, the colony will move to a cleaner environment. Likewise, the colony makes decisions about where to forage, when to send out new queens, when to go to war against a competing colony and when to make peace, and so on. These decisions are far beyond the brainpower of any single ant, even the queen. The queen, despite her title, is little more than an ovary; she is no smarter than her workers and is invested with no unique powers. But, then, how does a colony of ants think?

So far, we've seen the orderly progression of biological networks from the molecular level (cytoplasmic networks) to the cellular level (bacterial colonies and multicellular organisms). Can we push the concept further by assigning network architecture to ensembles of multicellular creatures such as an insect colony—or even the Internet? I think we can, and I'm hardly being original in this regard.

Long before there was a Turing, a von Neumann, or a Hopfield, all the way back in the year 1911, entomologist and classical scholar William Morton Wheeler wrote an influential essay entitled "The ant colony as an organism," in which he argued that insect colonies aren't mere *analogues* of whole organisms, they *are* whole organisms. Later, in 1928, Wheeler began referring to insect colonies as "superorganisms" to differentiate them from "metazoan" organisms like our own bodies (he was fond of neologisms).

Edward Wilson later summarized the similarities between cellular organisms and ant superorganisms as follows:

1. *The ant colony behaves as a single entity, like multicellular organisms;*
2. *The colony displays idiosyncrasies in behavior, size and structure that are peculiar to a given species and other idiosyncrasies that distinguish it from other colonies of the same species;*
3. *It undergoes cycles of growth and reproduction that are clearly adaptive; and,*
4. *It is differentiated into "germ plasm" (queens and males) and "soma" (infertile workers).*

(Adapted from E. O. Wilson's *The Insect Societies,* 1971)

These characteristics apply chiefly to so-called eusocial insects— ants and some bees. Eusocial insects have at least two "castes" (fertile parents and infertile workers), exhibit generational overlap within one colony, and share care for a common brood. Other types of social insects share in the care of the young or join forces to capture prey, but they lack the clear queen/worker differentiation found in eusocial colonies.

After Wheeler, others explored the superorganism concept, including Jean-Arcady Meyer, who in 1966 prophetically compared workers in an insect colony to neurons in the brain, and Alfred Emerson, who viewed the superorganism as a useful framework for studying social behavior. Wilson himself, however, dismissed the idea as inherently untestable: "The superorganism concept faded not because it was wrong but because it no longer seemed relevant.... The concept offers no techniques, measurements or even definitions by which the intricate phenomena in ge-

netics, behavior and physiology can be unraveled." (Wilson, 1971)
I have not seen anything in his subsequent writings to suggest that
he has changed his mind, although in his Pulitzer Prize–winning
The Ants (coauthored with Bert Hölldobler), published in 1990,
Wilson's ideas begin to sound suspiciously connectionist:

> *In theory at least, the organization of the colony can be described*
> *as the matrix of interaction of the members of the colony both*
> *within and across castes, subject to the constraints of ergonomic*
> *gaps and limited memory. Some of the workers will interact at*
> *very frequent intervals, other seldom. Most of the interactions*
> *will be cooperative and productive for the colony whereas others*
> *will consist mostly of interference and reduce production. The er-*
> *gonomic matrix presumably evolves toward higher fitness states*
> *by the genetic alteration of the relative frequencies and behav-*
> *ioral patterns of the castes and the details of the interactions. The*
> *matrix is therefore the scaffolding of the superorganism.*
>
> (Hölldobler and Wilson, *The Ants*, 1990)

The "ergonomic matrix" mentioned above is equivalent to a
connection weight matrix in that it defines the "matrix of interaction" among workers in the colony. Here, Wilson borrows the
word *ergonomic* from human sociology, where it is used to describe
the distribution of labor within a society.

Ant colonies contain up to several million individuals. Does
this make them huge, complex networks capable of great works of
emergent intelligence? Huge, yes. But complex? Well, maybe not;
ant colonies can contain millions of ants, but they don't necessarily contain millions of *neurons* in the formal sense. True, a solitary
ant is a complex, nonlinear I/O device, but individual ants don't
develop the type of unique interactions with other ants that can

drive a network. Individual ants communicate with one another, but experimental observation suggests that they are incapable of forming the social cliques needed to create complex E surfaces. In ant networks, I believe that the neurons aren't the ants themselves but the colony's castes and subcastes. The activation state of a caste would be defined by its size (a large caste would be "on," a small caste "off"), and the weighted connections would be defined as they would be in any party network, namely, as the statistical cross-section between any two castes.

In an earlier chapter, I noted that weights in very large party networks are statistical entities. Immune "neurons" are more likely to be *clones* of genetically identical lymphocytes than individual lymphocytes. Likewise, the ant neuron may be a *caste* of ants and not the ant itself. The true nature of the ant neuron (and the immune neuron, for that matter) awaits a great deal of additional experimental and theoretical work. Nevertheless, we know that for any interaction to be considered a network connection, it has to be trainable; otherwise, intelligent behavior can't arise. Thus, if we want to know how a given network defines its connection weights, we need only examine which of its internal interactions are influenced during the learning process. In ants, the connections between individual insects don't vary much, whereas the connections between *groups* of ants do vary according to environmental conditions. We can reasonably conclude that in an ant network, the trainable connections occur between large numbers of ants, not between isolated ants.

Because of the centralized nature of ant reproduction, worker ants (and the connections between ant groups) are genetically encoded in the queen's genome, just as enzymes and their mutual connections in a cytoplasm are genetically encoded in the cell's nucleus. Studies of the long-term evolution of social insects sug-

gest that colonies adapt over time by altering the size, characteristics, and interactions among castes and subcastes. In other words,
they evolve by changing their "ergonomic scaffolding" among
groups, not by modifying connections at the level of individual
insects.

Several mathematical models of social insects have been published since 1990. For example, entomologist Deborah Gordon,
mathematical biologist Brian Goodwin, and physicist Lynn
Trainor built a network model using ants as McCulloch-Pitts neurons and found that they could predict the size and interactions of
four worker subcastes: midden workers (garbage handlers), pa-
trollers, nest maintenance workers, and foragers. Even though
each ant was considered a neuron, the model deals primarily with
caste behavior. Their later models dealing with interaction proba-
bilities prove even more interesting, suggesting that castes, not individual ants, are the real neurons. For those who would read too
much into the emergent IQ of ants and other eusocial insects, this
is bad news; colonies are networks consisting of only a dozen or so
"neuronal" castes at best. Although it's difficult to precisely determine the number of subgroups within a colony, it certainly isn't in
the millions or even the hundreds. Castes can be defined in different ways: physically (workers, soldiers, and queens all have unique
anatomies); temporally (ants may shift roles as they age); and by
assigned task (foragers, midden workers, and so on). However
they are defined, the number of castes is small, in the single or
double digits. Colonies aren't equipped for solving Fermat's last
theorem, but they get the job done nonetheless. Every network is
only as smart as it has to be—intelligence isn't free.

Eric Bonabeau, Marco Dorigo, and Guy Theraulaz, working
at the Santa Fe Institute, devised an "ant optimization algorithm"
based on ant colony behavior in the late 1990s. These authors

applied a new name to the emergent computational power of ants: *swarm intelligence.* In their model, the connections among ants were defined by the chemical messengers, called pheromones, ants use to mark their trails. We know that ant colonies can solve certain optimization problems collectively, such as the shortest route to a food source; the Santa Fe approach explains this behavior quite well. In a paper published in *Nature,* the authors tried to distinguish swarm intelligence from the established discipline of network theory by pointing out four properties exhibited by swarms but not (allegedly) by networks:

1. swarm intelligence relies on the dynamic connectivity of ants (though, I would point out, so does a network);
2. swarm intelligence uses "mobile units" (but so do party networks);
3 swarm intelligence relies on feedback from the environment (what intelligent system doesn't?); and, finally,
4. swarm intelligence relies on pheromone evaporation to weaken connections and facilitate optimization (but, as I've discussed, networks must include a way of weakening, or decaying, connection weights; otherwise, unbridled Hebbian learning would soon saturate the weights).

To me, swarm intelligence is indistinguishable from a conventional network made of mobile neurons. In their book *Swarm Intelligence,* Bonabeau, Dorigo, and Theraulaz concede that the similarities between swarm intelligence and standard network theory are "overwhelming." Terminology aside, their work remains an original and important contribution to the understanding of living connectionism.

Bonabeau and his colleagues apparently believed that "networks" can contain only immobile neurons, hence their application of a new name—swarm intelligence—to ensembles of mobile neurons, including social insect colonies. In fact, they've merely come up with another formulation of network computation. Other researchers have also developed "new" theories of group behavior that aren't so new after all.

Despite their simplicity—perhaps *because* of their simplicity—ant colonies may be an ideal laboratory for dissecting biological network behavior. They have few neurons, use easily defined interactions, and operate in full view of the naked eye. In addition, insect species manifest a complete spectrum of social cooperation, from individualistic behavior to communal behavior to full-fledged eusocial behavior, and a comparative study of insect population dynamics might provide insight into how biological networks evolved in the first place (an excellent overview of this subject can be found in *The Evolution of Social Behavior in Insects and Arachnids,* edited by Jae Choe and Bernard Crespi). Extracting similar insights from a comparative study of animal brains would be difficult, mainly because we have no neurological equivalent of nonsocial insects. All animal neurons participate in brain networks, but not all insects participate in eusocial networks. Consequently, we can use existing insect species to trace the evolution of their eusocial behavior, but we can't do likewise for brain behavior.

Ant colonies manifest both genetic and non-genetic learning. The basic nature of colonies evolves over time secondary to mutation and selection—genetic learning. However, individual colonies also learn through post-genetic modifications of their caste structures and caste interactions. A simple example: if a human feeds an ant colony birdseed by hand each day, the colony will eventually stop deploying foragers; it will redistribute those ants to other tasks. The network turns the forager neuron off. Deborah Gordon, in her book *Ants at Work,* provides an even more sophisticated example. When two neighboring colonies forage for food in the same area, a colony can either fight or cede that territory to its rival. Young colonies (less than two years old) tend to avoid a fight, as do older colonies (more than five years old). Immature colonies may not have the workers to spare for a costly war, while older colonies have more established territories and fewer growth demands. Adolescent colonies, in an exponential phase of growth, must be more aggressive and may insist on foraging in disputed territories even when that activity generates mortal conflict. Like any organism, the colony changes with age. These behavioral changes are reflected in alterations of caste size (activation states) and statistical relationships (connection weights) among castes. Colonies start life with genetic learning and use it as a template for non-genetic learning, just like the brain. When Nature hits on a good idea—in this case, network architecture—it uses it again and again.

Undoubtedly, the most ambitious model of a superorganism ever proposed—one consisting of the entire planet—sprang from the fertile mind of James Lovelock in the 1970s. Following the suggestion of Nobel Prize–winning novelist William Golding, Lovelock called his superorganism Gaia, after the Greek goddess of the earth. Gaia unites both the organic and inorganic worlds into one giant network. Gaia theory became something of a New Age cult shortly after its inception, although anyone familiar with Lovelock's work knows that it's hardly the stuff of crystal-worshiping astrologers. On the contrary, Lovelock has advanced carefully reasoned, highly technical arguments to support his belief that the physical earth—oceans, rocks, atmosphere, and all—has been molded by life to suit its needs. Life didn't adapt to a fixed environment, says Lovelock. Instead, the physical environment and life co-evolved together, as a single, interconnected unit.

Consider the atmosphere. Before the advent of life, earth's air consisted almost entirely of carbon dioxide spewed from volcanic fissures, with a trace of methane thrown in for good measure. Gaseous oxygen was virtually nonexistent and, because of the "greenhouse effect" of carbon dioxide, the earth's surface stayed hot enough to burn paper. Today, the atmosphere is chiefly nitrogen with a healthy fraction of oxygen (21%) and almost no carbon dioxide (0.03%); this paucity of greenhouse CO_2 means a more comfortable (for life) average temperature. More important, the present abundance of oxygen keeps nearly every free element in an oxidized state. The oxidized state of hydrogen—water—is particularly critical for living processes. If a planet's atmosphere lacks free oxygen, all of its hydrogen—the lightest element—will eventually drift into outer space, dooming it to permanent aridity; without hydrogen, there can be no water and no life.

Biochemically, the dominant living systems on earth (photosynthesizing plants) extract atmospheric carbon dioxide, split it into carbon and oxygen, then release free oxygen into the air while at the same time "burying" gaseous carbon into vast reservoirs of solid coal and liquid oil. (Incidentally, we are now reversing the process by burning fossil fuels, thereby converting buried carbon back into atmospheric carbon dioxide, but I will not address the issue of global warming here.) Earth continues to spew carbon dioxide from its interior and life perpetually buries solid carbon while pumping oxygen into the air. The atmosphere we breathe now is not a "natural" environment, but one engineered by the biosphere over billions of years. Life doesn't thrive on earth because it has a cool, oxygen-rich environment; life itself cools the earth and keeps it oxygen-rich. This is the central thesis of

Gaian theory: life molds the earth, not the other way around, or, more correctly, the physical earth and living earth adapt to one another. Like two partners forced to change their individual ways in order to make a more successful marriage, earth and life have become one. Gaia represents their harmonious union, forged over eons of marital compromise.

Lovelock showed that life's effect on other elemental systems has been equally profound. For example, shellfish bury soluble calcium in their protective shells. As clams and other shellfish die and sink to the ocean's bottom, they precipitate soluble calcium to the ocean floor, where the great weight of the sea compresses it into limestone. Although limestone can arise inorganically, life accelerates the deposition of solid calcium on the ocean floor. The gargantuan accretion of limestone and other heavy calcium salts weakens and buckles the earth's crust, causing the tectonic movements that give rise to earthquakes, tidal waves, and mountain ranges. Even the weather comes under life's control. Rainforests manipulate their ambient humidity; they make rain where they live, they don't just live where it rains. Living things not only pump water into the atmosphere, they also produce substances that enhance droplet formation in clouds and induce precipitation.

Ocean algae secrete chemicals that reduce wave formation, and because of recent reductions in North Atlantic sea life, the average wave height has grown and the Eastern seaboard has been subjected to increased pounding and erosion. Microbial life may also directly contribute to rock weathering; geological processes, once thought to be purely physical phenomena, have a strong organic component. Lovelock likens the world to a great redwood. Most of the tree's massive bulk consists of lifeless cellulose and bark with only a thin rim of living plant material remaining, yet the tree can't be separated into living and dead components. Both the tree and

the earth are one living entity, a thin film of metabolic activity supported by a massive physical matrix of its own construction.

Critics of Gaia charge that Lovelock's theory is too teleological, because it likens the biosphere to some sentient being capable
of consciously manipulating its environment. Mainstream evolutionary theory considers evolution to be a blind process, possessing neither insight nor forethought. The notion that Gaia, the
Earth Mother, might be a conscious entity further fanned the New
Age flames and made the theory sound kooky. Looking back, I find
it fascinating to read Lovelock's original work and study how he
answered this criticism. He eventually resorted to computer modeling, as I'll discuss shortly.

Neither Lovelock nor his critics had a clear understanding of
the concept of emergent behavior at that time. Remember, this was
the 1970s, a time when few people owned computers and even
fewer knew anything about AI or the then-infant field of connectionism. In those unenlightened days, Lovelock had to come up
with his own way of showing how greedy, nonsentient beings might
cooperate to achieve emergent intelligence without the need for invoking an external godlike agency. Gaia's teleological flavor really
comes from network theory, not deism, but lacking a true connectionist framework, Lovelock turned to Daisyworld.

Daisyworld is a computer model created in 1982 by Lovelock
and his colleague Andrew Watson. In its simplest inception, the
planet known as Daisyworld contains only daisies. The daisy populations differ solely in the shade, or *albedo*, of their leaves. Albedo
measures the amount of sunlight the flower absorbs. A white
flower, for example, might have an albedo of 0.1 (it absorbs 10% of
sunlight and reflects 90%); a black flower, an albedo of 0.9 (90%
of sunlight absorbed, 10% reflected). Gray flowers would have
albedos somewhere between these extremes.

When life started on earth about 3.8 billion years ago, the sun's output was about 30 percent less than it is today. Several billions of years into the future, the sun will emit 30% more radiant energy than it does now (and we'll all need sunblock with an SPF rating of five trillion). Lovelock gave his computerized Daisyworld an evolving sun too, starting it out as a cold star and allowing it to warm with age. At the beginning of life on Daisyworld, there are just two daisy species, one light-colored (an albedo of 0.2) and the other dark (an albedo of 0.7). The bare ground has a gray shade of 0.4. The two species compete for resources, including water and nutrients. Lovelock used equations from theoretical ecology to model the growth of competing plant populations and set off to explore how the two species would handle a mutual threat: rising solar heat.

Initially, the planet is randomly (and sparsely) populated, with an equal distribution of light and dark daisies. Both species prefer an air temperature of 20 degrees C and will die if exposed to temperature extremes. Because Daisyworld is very cold at first, natural selection favors the dark daisies, because they absorb more light (and more heat) and are able to keep their local environment slightly hotter. The lighter daisies reflect most sunlight and begin to freeze to death, and the dark daisies dominate the planet early in its history. Soon, however, the expanding population of dark daisies causes the planet's air temperature to rise and, as the atmospheric temperature nears 20 degrees C (the optimal temperature for plant life on Daisyworld), even the lighter daisies begin to survive, even thrive.

As the sun ages and grows hotter, lighter daisies are actually favored, since they can reflect more heat than can darker daisies. During the life of the sun, the daisies' light/dark ratio rises; remarkably, the changing ratio manages to keep the temperature a constant 20 degrees C, even though solar energy continues to rise

with time. Eventually, on the hottest days of the dying sun, all
dark daisies are cooked and only lighter daisies persist. When the
sunlight overwhelms the ability of light daisies to reflect heat
away, they too die. After all daisies have expired, the temperature
rises linearly with solar age, and Daisyworld spends its golden
years as a glowing ember. A graphic depiction of Daisyworld appears below:

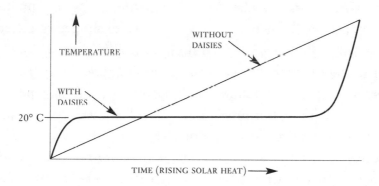

(Adapted from Lovelock, *Ages of Gaia*, 1988)

As the ratio of light to dark daisies evolves, Daisyworld shifts
its overall planetary albedo to keep the air temperature constant for
a large fraction of the solar life span. If we were to omit the daisies,
the temperature on a barren Daisyworld would rise linearly from
the birth of the sun until its death, and the lifeless planet would
spend only a small amount of time in the survivable temperature
range. Computer simulations of Daisyworld show that two daisy
species can maintain their ambient temperature in a survivable
range by using Darwinian selection in the face of an escalating environmental threat (rising solar heat). Writes Lovelock:

In Daisyworld, one property of the global environment, tempera-
ture, was shown to be regulated effectively, over a wide range
of solar luminosity, by an imaginary planetary biota without

invoking foresight or planning. This is a definite rebuttal of the accusation that the Gaia hypothesis is teleological, and so far it remains unchallenged.

The Ages of Gaia, 1988

Lovelock expanded his model to include gray daisies, and even daisy-eating rabbits and rabbit-eating foxes. The results were the same: a world populated with competing species can produce a buffered environment mutually beneficial to all species, a sort of global altruism (Yang) derived from local selfishness (Um). Dark daisies, by trapping the heat of a young sun, raise the global temperature, an act which unwittingly favors its lighter competitors. In the short term, this proves detrimental to dark daisies: they're forced to yield a greater portion of Daisyworld's limited resources to light daisies as the air temperature rises. In the long run, though, this act of faux altruism actually *benefits* dark daisies. If the light daisies were all left to freeze to death at the start of creation, the reign of dark daisies would be complete—but brief. With no light daisies to reflect away rising solar radiation, the amount of time Daisyworld spends at a survivable temperature would be severely curtailed. Likewise, in the twilight of Daisyworld, light daisies help keep dark daisies alive by keeping the temperature low. Neither species has any intention of being altruistic, and neither has the intelligence to foresee how its short-term altruism benefits it long-term. The emergent behavior of Daisyworld—the maintenance of global temperature homeostasis—arises emergently from a local, selfish competition among "unintelligent" flowers.

Daisyworld, like the Hopfield student network and the Bohr atom, is a useful preliminary model of immense historical and educational value, but we can't take it too seriously today. In truth, the two-daisy model isn't all that much different from a chemical buffer in which competing ions manage to keep an aqueous solution at a constant pH. Nevertheless, we can identify some basic network concepts in the Daisyworld model:

1. the neuron—the populations of daisies;
2. the connection weights—there are two connections on Daisyworld: the albedo (the daisy–sun connection) and the competitive interaction between the two daisy populations (the daisy–daisy connection); and, finally,
3. a single attractor basin—the ideal temperature, 20 degrees C.

Daisyworld is a party network; the connections between daisy populations are statistically defined. Yet daisies aren't "mobile" like ants or people, so how can they form a party network? In fact, the daisies *are* mobile; flowers slowly spread across the face of Daisyworld and meet their competition head on. Plants aren't very fast and they don't have legs, but they do move, aided by wind, water, and animals. (Plants obey the same network laws as animals, albeit on a different time scale.) The input to the Daisyworld network is solar light, and its output is atmospheric heat.

We can equate the activation state of daisy populations with its size. A small population represents an "off" daisy neuron and a large population, an "on" neuron. As Daisyworld evolves, the dark-albedo neuron goes from on to off and the light-albedo neuron switches from off to on. This shift in neuronal activation alters the planet's connection to the aging sun and keeps the air temperature inside its attractor basin at 20 degrees C.

There is only one thing missing from Daisyworld, but it's a huge thing: learning. The model's designers incorporated the optimal conditions for daisy growth and the details of daisy competition in their ecological equations. In effect, Lovelock and Watson engineered Daisyworld's attractor basin by hand. Daisyworld, like any nonlinear dynamical system, excels at finding its attractor, but this planet possesses no way of creating or modifying its attractor landscape in response to changing environmental needs. In this regard, Daisyworld again resembles a chemical buffer: the chemist designs the buffer to maintain a certain pH; the buffer does the rest. After he created an attractor valley for his Daisyworld network, Lovelock was pleased to find that the system promptly rolled into it and stayed there. Ironically, Lovelock and Watson proved just the opposite of what they had hoped: Daisyworld worked because an external agency designed it to work—and they were that agency. Lovelock and Watson became the gods of Daisyworld. On a real planet, the albedos would not stay fixed but would also evolve; I have no doubt that a completely mirrored daisy with an albedo of almost zero would certainly arise on a very old Daisyworld, extending its life span even further.

I don't mean this as a criticism of either Lovelock, for whose work I have great respect, or his Gaian hypothesis. The weaknesses of Daisyworld don't detract from the fundamental truth of Gaian theory, any more than the weaknesses of the Bohr atom nullify the truth of quantum mechanics. Lovelock's chief contribution to ecological network theory was his inclusion of *physical* neurons (the atmosphere, the ocean) in the *organic* network. His model affords us a brief glimpse of a unified ecological network spanning the whole globe. More work is needed to flesh out the Gaian model mathematically, but Lovelock's core argument must be correct. Ideas that beautiful usually are.

◎

Examples of superorganism networks are all around us. The United States Congress is an excellent example of party (no pun intended) connectionism. A representative or senator is, after all, a McCulloch-Pitts neuron, receiving input from many constituents, adding them, and using that sum to decide what binary state he or she will relay to the network at large (a yes or no vote on a given piece of legislation). Interestingly, the connectionist model of legislative bodies requires members to be entirely selfish. No single nerve cell can pretend to speak for the whole brain, nor can any senator presume to speak for the whole nation. The system works only when the senator from Pennsylvania votes for things that benefit only Pennsylvania. As we have seen, emergent altruism can arise from a competitive arena populated only by selfish individuals. We could extend the argument to economic systems by comparing free-market (connectionist) economies with centrally planned (serial) economies, but I'll leave that discussion for others.

No discussion of superorganisms would be complete without mentioning the Internet. That's about all I plan to do—mention it. The United States military supported the development of the Internet because it needed distributed communications networks resistant to injury (injury tolerance is one of the chief reasons living things prefer networks). The Internet is a scale-free network that is relatively immune to random knockouts, but how "intelligent" it may be is anyone's guess.

Some authors have gone overboard and attributed something akin to consciousness to the Internet and other global networks. This isn't as crazy as it sounds, given that the Internet now consists of many millions of connected humans and so approaches a vertebrate brain in size, if not in complexity. (A given Internet user can communicate with only a handful of other users at one time; a vertebrate neuron, on the other hand, can communicate with up to 80,000 other neurons simultaneously.) Nevertheless, we're left with some very basic questions: What is the Internet's input and output? What problem set does it collectively solve, or even address, and how are its weights trained to solve them? And as Turing would ask: how do we converse with it and prove that it's intelligent? I must side with Edward Wilson in this regard: the Internet as superorganism remains an untestable idea at this point in time.

I've tried to make sense out of the work of "global mind" theorists like Howard Bloom (*Global Brain*) and Pierre Levy (*Collective Intelligence*), but I find their work confusing and not very enlightening. These authors, like others devoted to the mystical concept of a global brain, don't bother with technical details, such as what constitutes a neuron or how the global connections might be trained. I'll admit that many of my ideas are highly speculative as well and may be proven wrong, but I've at least tried to provide a rational basis for them using known computational models.

Global-mind enthusiasts falsely believe that any sufficiently large collection of interconnected neurons must, by definition, make a "mind." While any large network can display *intelligence*, manifesting a conscious mind is another matter entirely. An ant hill meets all the criteria for an intelligent network—nonlinear neurons, weighted connections, modifications of connection weights during learning, and so on—but that doesn't put it in the same league as a brain. My young daughter competently blows notes through her bass clarinet in a pattern that meets all the criteria of music, but *The Magic Flute* it isn't. Spiderman comics and the Bible share a common ink-and-paper paradigm, but they are hardly equivalent literary works. Brains, ant colonies, immune systems, and ecosystems may share a common network technology, even a common intelligence, but they aren't identical. Different networks evolved to face different classes of problems, and to solve them over vastly disparate scales of time and space.

Consciousness could well be an evolutionary adaptation unique to large vertebrate brains; we simply don't know. The conscious state may represent a kind of weird phase transition that occurs when a network acquires sufficient hidden neurons to store an internal representation of itself. (A phase transition is an abrupt shift in a system's emergent properties—water turning to ice, for example.) We don't fully understand how informational capacity grows as a function of expanding network complexity, but let's assume for the moment that if the complexity of a network rises *linearly*, its capacity for storing internal data rises *exponentially*. In this scenario, a network that doubles in complexity would increase its storage capacity fourfold.

As the network grows more complex, the data it can store grow far faster than the data needed to describe its own construction, and at some point its capacity will exceed that needed to store the blueprints of its own design. For example, we have only about twice the genes of a fly. Thus, the number of genetic "instructions" needed to build a human brain may be only a little larger than the number needed to build a fly's brain, yet our brain has many orders of magnitude more data capacity than a fly's. The intelligence of a nervous system clearly rises *exponentially* relative to the instruction set needed to assemble it and so, at some point, a brain's capacity will exceed that needed to store—and to comprehend—its own design.

Unlike serial machines, network devices are built from repeating units (neurons) obeying a simple set of organizational rules, and it's conceivable that a network may become intelligent enough to comprehend itself. Serial machines, on the other hand, wind up chasing their tails in this regard. The more intelligent they become, the more incredibly complex they become. They may never reach the phase transition point where the intelligent machine's storage capacity equals or exceeds the amount of information needed to build it in the first place.

An identical phase transition must have occurred billions of years ago in enzyme networks, giving rise to life itself. Both life and consciousness may be comparable phase transitions in organic network architecture, where enzymatic and neural networks owned enough capacity to store their own blueprints. Global-brain advocates would have us believe that certain macro-networks, including the Internet and Gaia, have also reached or crossed their phase transition points for network self-awareness. They could be right, but I doubt it, and in the case of Gaian eco-networks, this assumption seems especially dubious. If we consider ant colonies as typical of superorganismal networks, we quickly realize that such networks fall far short of the data capacity needed for true self-awareness. At some point in the future, when we know enough about the quantitative aspects of biological networks, we might be able to prove that a network is conscious or alive on theoretical grounds alone. With due respect to Bloom and others, we haven't reached that point quite yet.

Interestingly, Lovelock and global-brain enthusiasts occupy opposite ends of the AI spectrum. In its first incarnation, Gaia theory actually *underestimates* global intelligence. Lovelock envisioned an ecological network dominated by feedback loops, a throwback to the old cybernetic concepts popular in the 1950s and 1960s. For all its intellectual beauty, the original Gaia theory viewed earth as a giant furnace equipped with thousands of thermostats. Again, this isn't a criticism of the concept itself, only of the outdated way in which it was first presented.

Global-brain enthusiasts, on the other hand, tend to *overestimate* the intelligence of large-scale networks, without providing any theoretical justification for their wild claims. In reality, such theoretical justification doesn't yet exist, because no one understands complex biological networks well enough to compare the intelligence of a rainforest to that of a chimpanzee. I stated earlier that any one species may be as intelligent as the next, when both genetic and non-genetic network learning are taken into account. The intelligence of Gaia is another matter. Species hone their intelligence in heated competition against one another, and that competition tends to keep intelligence equally balanced among competitors. But against what does Gaia compete? The only threats to Gaia come from astrophysical tumults—asteroid impacts, sudden shifts in planetary orbits with resulting climatic perturbations, massive volcanic eruptions, and (as on Daisyworld) shifting levels of solar heat. As Lovelock points out, Gaia has responded to all this and more with ease.

Intelligence is an expensive commodity and all things are as intelligent as they need to be, which, in the final analysis, means that all living things must be as intelligent as their competition, no more, no less. But Gaia's only competition comes from the physical universe. Gaia, the earth goddess, must therefore be as intelligent as the physical world that challenges her very existence. There's something very beautiful and moving in that concept. The biosphere has evolved to become the intellectual equal of the cosmic clockwork that spawned her.

If God did make something in His image, perhaps it was Gaia and not humankind.

9 MAGISTER LUDI

Hermann Hesse's Nobel Prize–winning novel *Das Glasperlenspiel* (The Glass Bead Game) tells the futuristic tale of Joseph Knecht, a common man who rose to become Magister Ludi, the "Master of the Game." Knecht held dominion over the "Glass Bead Game," the chief preoccupation of an elite twenty-fifth-century commune Hesse called Castalia. We'll ignore Hesse's marvelously complex plot and biting social commentary and focus instead on his fictional game.

As the name implies, the game was to be played with ornate glass beads, each imprinted with a symbol representing some deep truth about the universe. The player would create intricate patterns of glass beads to link different creative or scientific works. For example, the player could score points by showing a common bead pattern linking Beethoven's Third Symphony to Picasso's *Guernica,* or by illustrating a connection between quantum mechanics and Marxist economic theories. *Das Glasperlenspiel* assumes that all human knowledge, including science, art, music, and works of literature, shares a common set of irreducible building blocks. The only thing separating seemingly disparate intellectual creations—a symphony, a physical theory, a painting—is their unique pattern of ideogram beads.

The game doesn't really exist, of course, although many have tried to make a working version of Hesse's literary device. Nevertheless, the game's concept continues to intrigue us, for it has a basis in reality. The early twentieth century brought parallel revolutions in art, science, and psychiatry—all three began detaching themselves from the world of the tangible at the same time. Art went from realistic to abstract, physics from Newtonian to quantum, and psychiatry from the conscious to the subconscious. Abstract art, quantum mechanics, and Freudian psychiatry share the common bead of the unseen. We could go further and compare Einstein's theory of spatial curvature (more on this later) to Picasso's cubist reduction of the human form. In this instance, the shared bead is one of distorted geometry.

Common themes unite many diverse concepts and disciplines. Theoretical physics has been consumed with the search for a Theory of Everything, or TOE. The TOE, the theoretical physicist's Grail, is an (as yet) unknown way of including all physical phenomena under a single umbrella of laws. The quest for the TOE represents the *second* grandest of all glass bead games, and each new Magister Ludi, each Newton, Einstein, Feynman, and Gell-Mann, contributes new beads and new patterns, from quantum field theory to superstrings. But we're even luckier—in these pages we've been playing the grandest Glass Bead Game of them all, the Game of the Living World. In the previous chapters, I've provided a few of the beads we will need to make a TOE of biological systems. Needless to say, the final bead pattern of life eludes us, but we've made a good beginning.

Let's review what we know, beginning with the basic beads. They are:

1. the *neuron*, a nonlinear I/O device that can be linked to many others like it or to the exterior world;
2. the *activation rule*, which defines how a neuron converts its total input into a single output;
3. the *connection weight*, which regulates the flow of information between two neurons; and
4. the *learning rule*, which determines how the weights change in response to environmental influences.

We can add a fifth bead, the very important concept of:

5. *noise*, the random variations in neuronal activation states and connection weight values introduced by environmental agitation (or generated internally).

Noise can take many forms: mutations and other DNA missteps; electrical static in brain circuits; communication errors within ant colonies and human networks; direct thermal or radiation damage to enzymes, and so on. Rather than eliminate noise, life regulates it in a variety of ways: cells control the rate at which DNA damage is repaired; brains raise or lower their electrical signal-to-noise ratios; even immune systems appear to intentionally alter their noise levels. To understand the importance of noise, imagine a biological world without it. Evolution, for example, can't occur in the absence of genetic errors. Mutations are the marble that natural selection sculpts into improved life forms. In networks that use landscapes for solving problems, noise is a kind of raw material to be exploited. Perhaps that's why eons of evolution have yet to yield a "noise-free" biological system.

Systems that use attractor-based memories must have some way of raising and lowering the signal-to-noise ratio; otherwise they would stay trapped in some attractor basin forever. Consider an attractor basin of a certain depth, D, which has a little ball in the bottom of it. Let's now shake this basin as we would shake a mixing bowl containing a marble. The little ball begins hopping about and, if we define the average height of the hops as H, we'll see that the size of H depends on how vigorously we shake the basin: the harder we shake, the higher the ball tends to hop and H begins to rise. As H approaches D, the depth of the basin, we run a very large risk of shaking the marble out of the bowl entirely. In this case, D is the "signal" and H, the "noise"; D/H becomes this simple system's signal-to-noise ratio, which we control entirely by our shaking. If we shake the D/H ratio down to 1.0, the ball will likely exit the attractor basin. Likewise, *when living networks lower their signal-to-noise ratio, they're intentionally trying to "shake" themselves out of their present attractors.*

Why in the world would they want to do this? Because their present state may be wrong for the circumstances at hand. How many times have we tried to recall a melody or a person's name, only to have the wrong melody or name keep popping into our heads? We've fallen into an attractor all right, but not the correct attractor, so we must "shake" ourselves out and try rolling the little ball around the landscape again. A bacterial colony, when confronted with a new antibiotic, realizes that it's in the wrong attractor and begins shaking itself by increasing the rate of mutations, thereby raising the degree of genetic noise. It may not find a better attractor, but it has to pull up stakes and try. I consider the self-shaking process a kind of "thinking"; all living networks must be able to shake themselves spontaneously—that is, all living networks must think. The ever-changing nature of the real world means that no network, no matter how sophisticated or well trained, is in the right attractor all the time.

Metallurgists use a similar shaking process to switch metals from one attractor state to another. Recall that network architecture has its roots in physical systems, like the spin glass; steel is a network of iron crystals, and a piece of steel can exist in several different attractor states. The metallurgist can "shake" steel from one state (soft steel) into another (hard steel) by alternately heating and cooling the metal in a certain way, a process known as *annealing*. Heating agitates the iron crystals, increasing the noise in the steel network. In the cooling phase, the steel network's "little ball" quits hopping and rolls into another basin. If the heating and cooling are done in the correct fashion, the metallurgist can guide the network into the desired attractor basin: super-hard, annealed steel. Network theorists use an analogous process, called *simulated annealing,* to shake their networks from undesirable attractors into more desirable ones. By raising and lowering the signal-to-noise ratio in a prescribed fashion, computer scientists can "heat" and "cool" their software networks until they "anneal" into the correct configurations. However, annealing, whether done in computers or in metals, requires an outside agency (the metallurgist or programmer) to do the shaking, whereas living systems must shake themselves at the appropriate times.

The appropriate time is when the network's current state isn't getting the job done. We've explored two networks that shake themselves at *inappropriate* times: the schizophrenic brain and the cancerous cell. Both have extraordinarily low signal-to-noise ratios; they are shaking themselves continuously for no apparent reason. Anyone who has ever talked to a floridly psychotic person has been struck by that person's inability to settle into any coherent thought pattern. Here is a typical example of schizophrenic speech:

> *"I went to the mall, yes, the bear mauled me … do you like bears? I had a Teddy growing up, but I never liked Roosevelt, although my father did. I hate politicians, don't you? They're all so, so, so, oh, I do like to sew buttons, Red Buttons, you know, the actor, the* Poseidon Adventure *was so good, don't so you think …"*

This form of thinking, called flight of ideas, looks suspiciously like an over-shaken ball bouncing in and out of many attractor basins and never coming to rest. The cancerous cell, too, never comes to rest in any particular state. The hallmark of malignancy is its diversity; even in the same tumor mass, all cancer cells look a little bit different (as opposed to well-differentiated tissues, where all cells look identical). Cells from the same malignant tumor may even have different numbers of chromosomes.

I suspect that the perpetually reduced signal-to-noise ratio in cancer cells and schizophrenics may be secondary to the diseased state, not the cause of it. Cancer cells and schizophrenics share a detachment from the reality of the exterior world. In a multicellular organism, the cellular environment dictates the cell's differentiated state. Similarly, the social environment among humans dictates what does and does not constitute "appropriate" behavior. A cell behaves differently in a developing embryo than it does in an adult organism, and I behave differently in a men's locker room than I do in an operating room. In both instances, the local environment largely determines the correct attractor states of the cell or brain. Unfortunately, cancer cells and schizophrenic brains have trouble interfacing with their environments and generally ignore external cues. (In the case of cancer cells, this is secondary to defects in their signal processing hardware, their I/O neurons. The cause of the schizo-phrenic's detachment from reality remains unknown.) Conse-quently, "psychotic" networks have great difficulty finding the attractors appropriate for their respective environments.

And, even worse, networks *know* they're trapped in inappropri-ate states and so they lower their signal-to-noise ratios—the brain by altering its dopaminergic tone, and the cancer cell by switching off its DNA repair mechanisms—and start shaking themselves in the vain hope of locating a better attractor. When they fail (and they usually do, because their underlying defects aren't easily cor-rectable), they end up shaking themselves to pieces. If we spend time with a chronic, poorly controlled schizophrenic or gaze into the microscope at a paraffin section of malignant tissue, we'll read-ily see the pathetic wreckage of once-elegant networks desperately self-shaken into utter incoherence.

Often the shaking *does* come solely from the environment. In fact, every time the network is presented with a new problem, its stability is challenged; the act of seeking a new attractor becomes equivalent to reasoning out a solution. The great attractor jumps known as punctuated equilibrium were probably caused by envi-ronmental catastrophes of some sort. On Daisyworld, the flowers were eventually boiled out of their stable attractor. Since Daisy-world's creators gave it only one attractor, and no capacity to learn new ones, the ecosystem eventually collapsed. Whether noise is generated externally or internally, its volitional use by living net-works to jump themselves from bad attractors to better ones is a ubiquitous property of all intelligent life (and there is no un-intelligent life). In fact, self-shaking may be one of the defining characteristics of the living process itself. Whether we call the sculpting of noise "geological evolution" or "neurological thought," it's the same basic process—intelligence. By formulating both evo-lution and thought in common network terms, we are forced to reconsider the debate over the Divine Creator versus Dumb Luck Darwinism in a startlingly new light.

Life's use of randomness and noise has profound mathematical implications for the general theory of intelligent systems. Taner Edis, an associate professor of physics at Truman State University, points out that random noise allows living systems to break free of any single, predefined framework of "rules." In fact, as we've seen, such systems create their own rules. Some experts (including Roger Penrose) have asserted that any "rule-based" system will have trouble achieving true intelligence because of the famous incompleteness theorem of Gödel, which states that no self-contained system of rules can be internally consistent. Edis argues that a system that uses directed randomness (example: Darwinian networks) can escape Gödel's theorem. Space doesn't permit a detailed discussion of this controversy; I mention it only to emphasize that randomness and noise are more than useful for life—they are probably essential.

We also have a bead for:

6. the *attractor landscape* (or E surface)—a multidimensional surface defined by the matrix of connection weights and pockmarked with attractor basins representing network memories and solutions to various input/output problems.

And our growing bead set must also contain:

7. the *learning rate*—a measure of how fast the connection weights change during the learning process. The learning rate must be adapted to the situation. The disadvantages of slow learning are obvious, but fast learning can be just as bad. Rapid changes in the connection weights can cause the network to oscillate wildly without reaching a stable state.

The learning rate of biological networks appears to decline with age. Older people can't acquire languages and learn to play musical instruments as rapidly as children can, and immune systems don't respond to new pathogens as vigorously as they did in their youth. As previously noted, even ant colonies display different behaviors as they mature.

In brains, this decline appears linked to age-dependent chemical changes in synaptic receptors that make them less responsive to Hebbian influences. The age-dependent slowing of the learning rate isn't some cruel prank; it's a necessary adaptation. Young networks must acquire a massive amount of information quickly and so can afford to take more risks when modifying their connectivity. Older networks, with extensively configured E surfaces, don't want to rock the boat by making quick connection weight changes. In an old "Far Side" cartoon, a child asks to be excused from school because "my brain is full." In the case of well-trained E surfaces, this may indeed be the case; the E surface may be "full." Changes in one part of the surface ripple into other parts, and there comes a time in any aging network when change becomes destructive.

We now need to assemble the beads into a network. A complete network is equipped with input neurons, to gather information from the environment and relay it into the network; a "filter" of hidden neurons, to manipulate information internally; and output neurons, which convert the "filtered" data into external actions.

I refer to the set of hidden neurons as a filter because this is an accurate description of what a network does. The hidden neurons mimic the metal screen inside a pasta maker: we feed dough into the machine (input), grind it through the screen (hidden neurons), and out comes spaghetti, or angel hair, or fettuccine, or whatever else we choose to make (output). The geometry of the pasta machine's screen determines the resulting noodle output; likewise, the geometry of the E surface built into the connection weight matrix determines how environmental "dough" gets ground into different types of output "pasta." When we train a network, we change the pasta screen by altering the machine's internal landscape, thereby forcing the network to make different noodles from the same dough. A championship baseball player has a neuronal filter that can turn fastball dough into home run spaghetti. My own brain, on the other hand, has a filter only capable of turning fastball dough into strikeout spaghetti. Same dough, but different training, different filters—and very different pastas. Learning is equivalent to changing the innards of a pasta machine; it's that simple.

While we're on this subject, let's add the bead for:

8. *learning*, the act of altering connection weights within a living network.

Biological learning can be either (1) *genetic*, altering connection weights indirectly via the alteration of genes, or (2) *non-genetic*, altering connection weights directly in an existing network. We call genetic learning *evolution*, and non-genetic learning... well, we call it learning. Genetic learning is usually heritable (although the genetic learning in vertebrate immune systems isn't); non-genetic learning vanishes forever once the network dissipates, although there may be exceptions to this rule too.

All learning, indeed all intelligence, *occurs primarily via the application of some local learning rule, such as natural selection.* Immune learning, brain learning, and conventional genetic evolution employ similar computational methods, albeit over different time scales and using different hardware. The brain represents evolution modified into a high-speed, high-tech, portable form. Before genes, all learning was non-genetic; changes in connectivity among proteins required direct alteration of the proteins themselves, and there was no way to record the effect of these changes for posterity. After the advent of genes, most learning became genetic; network connections were modified at the level of the genome, thus allowing network learning to be passed on to later generations. With the advent of nervous systems, non-genetic brain learning again rose in importance. (The development of written language reintroduced a "genetic" form of learning into human networks, in that the connections among humans can now be recorded in a "heritable" format—on paper, for example—for future generations.)

Different species parse their methods of learning differently. Single-cell organisms, multicellular plants, and simple invertebrates still favor genetic learning (evolution), while vertebrates favor non-genetic learning. The two forms of learning are, to a degree, mutually exclusive. Our large brains prohibit us from birthing large litters, and our non-genetic learning engines (brains) take decades to train, thus burdening us with long generational times as well. A large non-genetic learning capacity tends to reduce our capacity for genetic learning. We are non-genetic scholars and genetic dunces.

Genetic and non-genetic learning operate on different time scales, but one isn't "better" than the other. Because we traditionally view evolution as a blind and stupid process, we have grossly underestimated the intelligence of genetic learners (like bacteria) and grossly overestimated the intelligence of non-genetic learners (like us). In fact, given that intelligence is essential to life and carries a price tag in terms of energy resources, the total intelligence (genetic and non-genetic) of competitive species *must* be comparable; otherwise, the more intelligent species would soon dominate.

Although modern species may have a comparable intellectual level, I believe that the overall level of intelligence on the planet has increased with time. *Intelligence is itself an autocatalytic, nonlinear process.* The sophistication of biological networks, both on the small scale and the large, has risen dramatically over the eons. Technological advances (the first enzymes, the appearance of the genetic code, the advent of the nerve cell, the genesis of human speech) increased life's computational power, thereby fueling greater adaptability and spurring additional advances. This positive feedback turned the earth into a breeder reactor for intelligent systems wherein intelligence breeds intelligence. The living world we see today *is* the work of intelligence, the intelligence of Darwinism operating within a network paradigm that has its origins—as Hopfield showed—in the physical world. Some authors contend that life must have an intelligent designer, but they don't define "intelligence," nor do they consider that the intelligence of the designer may have arisen here on earth, from the living process itself.

Those who would point to our so-called dominance of this earth as evidence of our intellectual superiority, not so fast! We're *losing* our war against bacteria, parasites, cancer, and the AIDS virus because our non-genetic brains can't keep up with the genetic genius of these pathogens. Their weapons evolve faster than our brains can think of ways to combat them. Anyone who believes otherwise hasn't been paying attention. Our only hope lies in forming quasi-superorganism networks (like the U.S. National Institutes of Health) that amplify our individual intelligence. Our need to communicate and form communication networks may be our only way of compensating for a stupendous lack of genetic adaptability and, unless we get smarter in the non-genetic sense, the "stupid" parasites of the world may quickly make us their dish *du jour*. Selection pressures have forced us to be social creatures, because we've sunk our life savings into the non-genetic paradigm and there's no turning back now. Our creation of written language may be our way of thumbing our noses at the microbial world and other genetic savants, including cancer: "Nyah, nyah, now we can try out a ton of mutations, too, with pen and paper." Sadly, the genetic learners haven't been very impressed with our efforts so far.

Some mourn for the earth in the reign of Homo sapiens, but fear not. The majestic Gaian network has dealt with far worse than us. If we push the great ecological ball too hard, it will simply jump out of this steady state and roll into another. Of course, *we* may have no home in that new attractor basin, but the biosphere will thrive nonetheless. Humans are just another form of Gaian noise.

We love our present attractor basin, but history tells us that no system stays locked in one state forever. I wouldn't be sitting here today, enjoying my little plastic cup of chocolate pudding and hoping it doesn't drip on the keyboard, had the earth's eco-network not rolled out of its deep Precambrian attractor and spawned segmented monstrosities like myself. The cancer victim bemoans his fate—death at the hands of an aberrant attractor jump—but forgets that a countless series of such "mistakes" gave him his life and his complex network of a body in the first place. We condemn noise when it harms us, but noise created us all. And noise, like death, taxes, and the poor, will be with us always. The next attractor jump may be just around the corner…for you, for me, or even for the great Gaia (in which case, God help us all).

Finally, to complete our bead set, let's add just one more bead:

9. *chaos,* the element of unpredictability.

I don't use the word *chaos* here in the colloquial sense, but in the mathematical sense. Chaotic systems operate under a unique set of rules, and the study of chaos has emerged as a distinct scientific discipline (back in 1987, James Gleick published the definitive history of the subject). Chaotic systems, like all dynamic entities, still seek their attractors, but chaotic attractors are much different from any we've dealt with so far. Before we can explore chaos any deeper, we need to take a short detour. Chaos is a complex subject, but its application to networks may eventually have some role in explaining abstractions like free will and consciousness, and so it's an important topic.

Up until now, I've talked only about *point attractors,* stable states represented by a single point in space (example: the bottom of a landscape valley), but there are other types of attractors, such as *periodic attractors.* To understand the periodic attractor, consider a simple pendulum. According to Newtonian mechanics, the period of a pendulum (the time it takes to swing one arc length back and forth) is solely a function of the pendulum's length. A twelve-inch pendulum will swing back and forth in about one second. This period is inversely related to the pendulum's *frequency,* the number of "swings" per unit of time. The shorter the period, the higher the frequency; a pendulum with a period of one second will have a frequency of sixty swings per minute, while a pendulum with a period of 0.5 seconds will have a frequency of 120 swings per minute, and so on.

For the pendulum, *the frequency is a periodic attractor.* No matter how hard and fast we swing a foot-long pendulum at first, it will quickly settle back into its one-swing-per-second frequency and stay that way until its motion decays. The frequency is a stable "periodic" state for a swinging pendulum, just as the bottom of a valley is a stable "point" state for a rolling wagon, the only difference being that the pendulum reaches its stability *while still moving.* A periodic attractor is, therefore, a "dynamic" stable state. The stability of the pendulum's periodic attractor makes the pendulum an ideal mechanism to drive clocks. In fact, all clocks contain some sort of oscillator trapped in a periodic attractor. Atomic clocks use vibrating cesium ions, electronic wristwatches use vibrating quartz crystals, and spring watches use an escape wheel mechanism driven by a spring. Like the simple string pendulum, all of these oscillators have a periodic attractor defined by their physical construction. A clock's machinery merely translates the fixed frequency of some stable oscillator into the language of human time.

Living systems employ both point and periodic attractors. Not surprisingly, biological periodic attractors often serve the same purpose in life as they do in clocks: keeping time. Scientists have discovered two clocklike gene systems in the Drosophila fruit fly, which they colorfully named *period* and *timeless.* The levels of the proteins encoded by these clock genes—per and tim—form an oscillator with a twenty-four-hour periodic attractor. Such oscillators are said to be circadian (from *circa diem,* "about a day"). Needless to say, per and tim are hardly unique to the fruit fly. Even single-cell creatures contain some form of circadian protein clock that drives their life cycles.

All networks contain some type of clock mechanism. In electronic models of networks, the clock sets the rate at which activation states are updated and connection weight changes implemented. Depending on the network, the system can be updated synchronously (all changes made at the same time) or asynchronously (each neuron and weight updated randomly, on its own schedule). Recall that our student Hopfield network was asynchronous: students looked at their meters and flipped their switches every few minutes at their own pace. The brain possesses a clock (near the pituitary gland), and even the immune system obeys the hands of time. A major lymphoid organ, the thymus, slowly vanishes during childhood according to a preordained schedule.

One last point: McCulloch and Pitts likened the nerve cell to a simple on-off switch, when, in reality, it's more like a pendulum. True, nerve cells go on and off like switches, but they do so in regular cycles. The "firing rate" of a brain neuron—the number of action potentials it releases per second—can be compared to the periodic attractor of an oscillator. Simple oscillators, like simple chemical reactions, are mono-stable. The string pendulum has one, and only one, periodic attractor, as defined by its length.

More complex oscillators, like the nerve cell, can gravitate to different periodic attractors, somewhat like a musical instrument gravitating to different harmonic frequencies. The nerve cell not only has two point attractors (on and off), but seeks multiple periodic attractors as well. *Nerve cell firing frequencies govern information flow in nerve networks.* In other words, the brain's network relies more on periodic attractors than on point attractors. This may hold for enzymatic networks as well, given that many enzyme systems oscillate like pendulums too.

Biological systems aren't really assemblies of snapping switches, but coordinated orchestras of periodic neurons creating the melody of life. Hesse begins *Das Glasperlenspiel* with a fictional history of the Glass Bead Game. As it turns out, the game was first conceived as a shorthand for musical theory and was later expanded to include other disciplines. In the final analysis, all knowledge reduces to patterns of harmonic frequencies.

Life is music. Music is itself an attractor-based discipline, in that there are no "smooth" transitions between notes; the octave is a punctuated equilibrium, with discontinuous jumps between pleasing tones and half-tones. Chords consist of defined frequency patterns. We can lump any group of frequencies together and obtain an aesthetic harmony. The brain's concept of musicality, indeed its concept of beauty in any shape or form, flows from its interior design and a love of patterned data.

Now, back to chaos.

Suppose we take our string pendulum and whack it at regular intervals. If the frequency of our whacking equals the natural frequency of the pendulum, we will simply reinforce its periodic behavior. But what if we whack it at some other rate? We will then disrupt its attempt to reach its periodic attractor. The pendulum will behave erratically and, at some point, switch into a chaotic form of swinging. We've all had the experience of being pushed on a playground swing by a playmate. If the playmate pushes in synchrony with our swinging, we maintain a regular period and our motion doesn't decay. If our playmate's pushing is out of synch with our swinging period, our motion may degrade into a chaotic pattern.

Not all irregular behavior is chaotic, however. Chaos is a special state that can also be defined as an attractor—a *chaotic attractor.*

The essence of chaotic movement is that it's essentially unpredictable, even though it occurs in completely deterministic systems, like pendulums. Before the discovery of mathematical chaos theory, scientists believed that only quantum systems, like the atom, were unpredictable. According to the Heisenberg Uncertainty Principle, subatomic systems can be so perturbed by the act of measurement that no observer can say with certainty where a subatomic particle is located and exactly what it's doing at any given point in time. To measure something, we have to "see" it, and to see it, we have to shine some form of radiation on it. To "see" an electron, I have to hit it with a particle of light (a photon). Because the electron is so minuscule, the collision between photon and electron isn't exactly inconsequential. Thus, the act of measuring very small things can be affected by the measurement process itself. The Uncertainty Principle states that we can't measure the subatomic world with infinite precision because the act of measuring always injects a certain degree of error into the equation (noise, as I said earlier, is everywhere).

However, macroscopic systems, like billiard balls and clock pendulums, are much too large to obey the Uncertainty Principle—photons have a negligible effect on them—and so they should be deterministic. In other words, if we know the *precise* initial conditions of a macroscopic system (at time $t = zero$), we should be able to predict with certainty how the system evolves with time, what it will be doing at time $t = 1$ *minute*, $t = $ *ten minutes,* or whatever. A billiard expert depends on her balls and table behaving in a deterministic fashion. If she hits the cue ball a certain way (at $t = zero$), the ball will roll over there and strike a second ball, which will roll where she wants it to, and so on. She can predict that the table, when subjected to the initial condition she defines—the cue stick striking the cue ball—will evolve in a deterministic way and end up pocketing the eight ball two seconds later. If she were hitting electrons with a photon stick, she wouldn't win many titles.

Of course, it's not possible in the real world to exactly know the initial conditions of any system. This is usually not a problem, because in a typical macroscopic system, any error in estimating the initial conditions only expands *linearly* as the system evolves. If the billiard expert miscalculates the angle of her cue strike by just a millionth of a degree, the discrepancy between where she *thinks* the ball will go and where it *actually* goes will linearly increase the farther the ball travels. If she's aiming for the moon, a millionth-of-a-degree error would be lethal, but on a pool table, it's trivial. She'll still sink that eight ball.

A chaotic system differs from a pool table in that any error in estimating the initial conditions is transmitted *exponentially* to the evolving system, not linearly. Let's analyze the behavior of a chaotic and a nonchaotic cue ball. Again, assume that the expert misjudges the cue strike by a minuscule amount. In the normal, nonchaotic system, the difference between the predicted and actual ball trajectories increases linearly as the ball travels. After the ball travels a meter, suppose the error in its flight path is one centimeter—that is, the difference between where the player thinks she's hitting it and where the ball actually goes is one centimeter. After a two-meter roll, the error will be two centimeters, and after a three-meter roll, it will be three centimeters. Not an insignificant deviation, yet still manageable. The player won't win many games this way, but the ball won't fly off the table.

But suppose the expert makes the same initial error on a *chaotic* pool table. After a one-meter roll the error becomes ten centimeters, then a one-hundred-centimeter error at two meters, and a staggering ten meters after rolling only a mere three meters! The slightest miscue, and the man in the front row takes a ball to the nose.

In a chaotic system, the smallest initial errors grow so fast that they render the system essentially unpredictable, even though it's technically "deterministic." Nonchaotic systems are mildly affected by slight changes in the initial conditions, but chaotic systems are exquisitely dependent on such changes. This unusually severe dependence on initial conditions has been poetically referred to as the "butterfly effect"; if we consider weather a chaotic system (which it is, by the way—that's why the local weatherman can appear so unreliable), a butterfly flapping its wing in India may eventually trigger a hurricane in Florida. Slight perturbations rip through chaotic systems violently, and those systems soon assume the metaphysical vagueness of quantum systems. If chaotic systems are so unpredictable, what good are they? Fortunately, unpredictability doesn't always equal uselessness. Wall Street is chaotic, but it still serves a purpose.

We can't predict the exact location of an electron in a hydrogen atom, but we still know that it exists in a spherical statistical distribution around the central proton called the 1s orbital state. The 1s orbital is a quantum attractor for the trapped electron, and although the attractor doesn't define exactly where an electron is at any time, it does impose major limits on the electron's behavior. Likewise, a chaotic attractor sets limits on a chaotic system's behavior, although it doesn't dictate the exact behavior of that system (as opposed to point and periodic attractors, which *do* define exact behavior).

To understand the difference between the three types of attractors—point, periodic, and chaotic—think of a bee buzzing around a flower. The flower defines the general geometric location of the attractor. If the bee comes to rest on the flower, the flower becomes a point attractor. If the bee oscillates regularly around the flower, back and forth with a defined frequency, the flower defines a periodic attractor. However, if the bee buzzes unpredictably around the flower (as bees tend to do), the flower becomes a chaotic attractor. In the last case, we may not know exactly where the bee is in space at any one time (and, in a chaotic state, it will never be in the same place twice), but we can still say that the bee will be around the flower. We can still know something useful about a chaotic bee (the bee isn't just anywhere in the field), but we don't know as much about a chaotic bee as we do about point and periodic bees.

I recently said that a string pendulum has only one periodic attractor, and that's still true, but if we drive the pendulum externally in just the right way, we can force it into another, chaotic attractor. Thanks to the existence of chaos, even this simple device now has a fairly complicated landscape, with a point attractor (at complete rest), a periodic attractor (swinging regularly), and chaotic attractors (swinging irregularly). A chaotic pendulum's frequency hovers around a certain periodic frequency, just as the bee hovers around the flower.

We know that biological pendulums, including nerve cells and enzymatic oscillators, can be driven into chaotic attractors. Why is this important? Because chaos instills unpredictability in biological networks without having to invoke quantum mechanics, *à la* Roger Penrose. Quantum uncertainty may still play a role (particularly if we factor in quantum computation, an interesting discipline that is beyond the scope of the present book), but it isn't needed to explain the unpredictability of network behavior. Moreover, the use of chaos by biological systems further distinguishes them from conventional "Turing machine" computers that rely on point and periodic attractors. Computer networks employing chaotic neurons are still in their infancy, and computer science has a long way to go before it can duplicate the complexity of living networks. Will we someday be able to duplicate a brain with ten billion chaotic neurons and ten trillion sophisticated connection weights, as Ray Kurzweil suggests? Perhaps, but

I'll propose another rule: if we take the degree of complexity Kurzweil and other AI mavens ascribe to a living network, then multiply it by the number of water molecules in the ocean, we'll obtain a more accurate idea of how complicated biological networks truly are. Let's just say that I'm not putting in my order for a robot butler just yet.

Suppose two identical networks approach the same problem at the same time. If chaos is involved, they will each reach a slightly different solution. By embracing chaos, life harnessed yet another source of noise and another way to bring diversity into computation so that natural selection could choose among many possibilities, not just a few. The more we understand the nuances of natural intelligence, the more amazed we become. No aspect of nonlinear dynamics has been left unexploited. The religious concept of free will may well derive from chaos, not from Heisenberg. (If this is true, then even cells have free will—another tidbit for armchair theologians.)

There is a deep mathematical relationship between chaos and *fractals*. French mathematician Benoit Mandelbrot discovered fractals (short for fractional dimensions) in the late 1970s and early 1980s, about the same time that chaos theory evolved. The simplest example of a fractal object is a wadded piece of paper. When the paper is flat, it's two-dimensional, but when we crumple it into a ball, it begins to enter the third dimension. As we crumple it tighter and tighter, we will eventually drive all the air out of it and compress it into a three-dimensional ball of solid paper pulp. The intermediate state between a two-dimensional paper and a three-dimensional ball—the crumpled state—must have a dimension between two and three, between flat and solid. In other words, a crumpled piece of paper has a fractional dimensionality; it's a fractal object.

The complex surfaces of clouds, cauliflower, and mountains resemble crumpled paper and are also fractal objects. A jagged coastline, somewhere between a one-dimensional line and a two-dimensional plane, likewise exists in a fractional dimension. If we examine coastlines, mountains, cauliflower, and clouds closely, we'll observe an important property of fractal objects: they display *scale invariance*. A small part of a mountain (a rock, for example) tends to look like the whole mountain, just as a small part of a cloud, cauliflower, or a coastline resembles the whole cloud, cauliflower, or coastline. No matter the scale, the structure of a fractal object appears similar. Living networks are also fractal in that they display the same general architecture at vastly different size scales. Cytoplasmic networks look like brain networks look like eco-networks. Incidentally, there is a fourth type of attractor, called a strange attractor, that relates to chaotic behavior in fractional dimensions. We'll stop here before someone's brain explodes.

As we come to the end of this book, the question arises: was Edward Wilson right? Is the theory of living networks (super or otherwise) testable, or even useful? I believe the answer is yes on both counts. The network architecture of the brain has already been experimentally verified to a large extent, and encouraging work continues in the realm of enzyme networks, insect networks, genetic networks, and ecological (Gaian) networks. Curiously, although Jerne was awarded the Nobel Prize partly for his 1974 theory of immune networks, serious interest in these networks has largely evaporated in recent years. Ironically, this may be the one area with the most immediate experimental and therapeutic potential. Lymphocyte networks are among the easiest cellular networks to isolate and dissect, yet they remain largely unexplored. The reason may be a knowledge barrier; few immunologists understand mathematical network theory in any depth, and few network theorists bother to learn the intricacies of immunology.

I believe that the immune system is vastly more intelligent than we realize, but since we don't speak its language, we tend to deal with it only through violence. When we transplant a heart or kidney, we suppress organ rejection using toxic drugs like cyclosporine, effectively beating the immune system into submission. It doesn't have to be this way. Many years ago, I visited the Louvre in Paris and, like most Americans, my French consisted of speaking English very loudly. I mistakenly paid for my ticket with an insufficient amount of money, and the employees had to accost me physically in order to get me to pay the correct amount. Here we were, intelligent beings, forced to communicate with pushes and shoves simply because we couldn't speak each other's language. So it is with the immune system. We can't talk to it, so we kick it until it does what we want. A more thorough understanding of immune networks may allow us to speak to our immune systems as easily as Dr. Doolittle spoke to animals. ("You want me to accept this kidney? Why didn't you say so in the first place?") This may be a bit of an exaggeration, but we can certainly try harder to communicate intelligently with such a sophisticated machine.

In an earlier chapter, I discussed how the many different antibodies that form against a single, large antigen may express a common idiotypic marker. For example, if we immunize an animal with myoglobin (a large muscle protein), the animal will soon produce dozens of different antibodies, each directed at a different area of the protein. These antibodies possess different V-regions, but they share the same idiotypic marker. The idiotype labels these different antibodies according to "content"—they all react to myoglobin—much like a Dewey decimal code labels different books that deal with a common subject. Idiotype labeling, also known as the Oudin-Cazenave phenomenon, provides powerful evidence that the immune system uses some system-wide, content-addressable memory, much like the brain.

The brain tends to categorize memories according to temporal sequence, even when those memories have nothing else in common. A man might forever associate the smell of his Uncle Ned's aftershave with the Superbowl, for instance, because Uncle Ned once got him great tickets. Does the immune system do likewise? In infancy, we're vaccinated against diphtheria, tetanus, and pertussis simultaneously. Can the immune system remember this when we're later vaccinated against tetanus alone, say, by increasing responsiveness to the other diseases at the same time? Does the immune system associate tetanus and pertussis in the same way that the brain associates a certain aftershave with a sporting event? I don't believe any experiments have ever been done to answer this. The possibilities are endless. Lymphoid networks, like all biological networks, have noise (although the precise nature of this noise is beyond the scope of this discussion), and the role of immune cytokines in manipulating the immune system's signal-to-noise ratio is another fertile area for future investigation. But enough about immunology.

As Robert Jackson and others have suggested, the landscape concept could impact our view (and treatment) of malignancy. We can no longer consider cancer as a disease of a single cell. Even if the genetic mistakes leading to cancer take place within one cell, the malignant state has meaning only in the context of the community at large. A sociopath can't be sociopathic living out his life alone on a deserted island, since a sociopath is defined as someone who ignores societal rules. In the absence of a society, there are no societal rules. Malignancy represents a failure of landscape recall and could be reversed by redirecting cancerous cells to some phenotypic basin. As is the case with the immune system, we resort to violence against cancerous cells because we don't speak their language. We can't communicate on any serious level with misbehaving cells or with the bodily network that keeps them in their differentiated states, so we just kill them, or try to without success. Killing is a useful stopgap measure, perhaps, but not the safest, most elegant or definitive of solutions. Some progress has been made using "differentiating agents" to reverse the malignant phenotype in certain cancers, but this approach remains out of the mainstream for now.

One approach to cancer may involve simply elevating the cancer cell's signal-to-noise ratio. This is similar to what psychiatrists do when they prescribe dopaminergic drugs like chlorpromazine for schizophrenics. Chlorpromazine forces the schizophrenic brain to stop shaking and fall into some attractor basin, any attractor basin. It's sort of like musical chairs: chlorpromazine turns off the music and makes the brain take a seat somewhere. That attractor won't necessarily be optimal; even a treated schizophrenic may not manifest entirely "appropriate" behavior (although exactly what constitutes appropriate behavior seems difficult to know these days). But the treated behavior will still be more rational and coherent than the behavior manifested by the perpetually shaken state. Perhaps we can similarly "anneal" cancer, reassuring it that an inappropriate attractor may be better than no attractor at all.

Our lack of knowledge about ecological connectionism prompts oafish attempts to restore "ecological balance" as we define it. For example, we transplant predators from one area to another in order to reduce a certain population, only to make a

habitat go from bad to worse. We fret about the effects of pollu-
tion, global warming, the loss of a certain species of toad, acid
rain, and a host of other so-called ecological threats, yet our
knowledge about global networks is so poor that we have no accurate idea as to whether such problems pose a serious threat to
network integrity. The earth's ecosystem is probably a scale-free
network, meaning that it can resist most injuries...but not all.

I'm not suggesting that we ignore ecological threats. On the
contrary, an enhanced theoretical understanding of Gaian networks might reveal that some things, like global warming (if it
exists at all), are *more* of a threat than we think. One benefit of
comprehending how networks work is a better understanding of
how networks *fail*. We know, for instance, that scale-free networks
fail when their most highly connected neurons malfunction. Thus,
we might learn which enzyme pathways are critical to maintaining
a cell's non-cancerous state simply by identifying the most connected ones. Much research has been directed to network failure
by telecommunications companies and various military organiza-
tions, but such research is either proprietary or classified. This is
truly unfortunate, given how many of humankind's greatest ills,
including cancer, schizophrenia, and ecological disasters, come
from network failures.

We could also work a little harder to understand how brain
networks craft their signal-to-noise ratio to enhance or suppress
learning and creativity. Artists and writers have long used signal-
suppressing drugs like opium, marijuana, and alcohol to increase
their creative output. Could we harness this in a more socially ac-
ceptable (and medically controllable) way? Can we develop some
means of stabilizing the learning rate in aging brains, at least tran-
siently? Maybe we could even develop a way of restoring the
learning rate to that of a toddler, so I might learn French and
avoid being throttled by angry Parisian docents.

Finally, we shouldn't ignore the lessons of connectionism from a social and political standpoint. No single neuron can comprehend the needs of the whole network, and no single brain, no matter how benevolent and intelligent, can direct an entire society, government, or economy. The old Soviet Union tried to replace the intellect of the masses with a handful of brains, and their farmers routinely received shipments of snowmobiles in place of tractors.

Earlier, I made a brief reference to Einstein's use of spatial curvature. In his General Theory of Relativity, which is really a theory of gravitation, Einstein replaced gravitational force with the concept of the *geodesic,* the shortest line between two points in space. Note that I say the *shortest* line, not the *straightest* line, because a geodesic may not always be straight in the Euclidean sense.

Suppose I want to go from Pittsburgh to Cleveland. I could go to Cincinnati first, but that wouldn't be the shortest route. Everyone can trace the shortest route between Cleveland and Pittsburgh on a globe, right? But that route isn't really the shortest one in a geometric sense, because a traveler must still follow the curvature of the earth. Technically, taking the shortest path between the two cities would require tunneling through the earth in order to circumvent its curvature.

If we're confined to a curved surface, the shortest path between two points may still be curved. In the General Theory of Relativity, Einstein discarded the idea of a gravitational force and replaced it with the idea of spatial curvature. The sun doesn't pull on the earth, said Einstein, but instead warps the space in which the earth must travel. The planets make quasi-elliptical paths around the sun because they are tracing their geodesics—their shortest paths—through a warped space, like marbles whirling inside a metal bowl. Einstein's odd idea has been experimentally confirmed many times.

Flush with the immense success of the General Theory of Relativity, Einstein sought to reduce all physical laws to a matter of geometry. He personally failed, but his ardent belief that all physical phenomena are just manifestations of some exotic, unseen geometry hasn't gone away. On the contrary, it has flourished. Modern physicists have gone a step further than Einstein and invented entirely new spaces to hold their entirely new geometries. Consider the example of *isospin space*.

Shortly after the neutron was discovered, physicists realized that the particle looked and acted much like a proton. A proton carries a positive charge and the neutron has no charge, that's true, but otherwise they seemed like the same particle. They're virtually identical in size and, within the confines of an atomic nucleus, behave in pretty much the same way. This prompted some theorists to propose that the neutron and the proton aren't just similar, but identical. The proton and the neutron are the same particle "viewed" from two different angles.

Assume for a moment that we're all two-dimensional creatures with no concept of a third dimension (assume also that the third dimension still exists). Along comes a quarter, a 3-D piece of U.S. currency. What would it look like to us flatworlders? Well, that would depend on the quarter's orientation in three dimensions with respect to our world; it might look like a thin rectangle (when viewed from the edge) or like a circle (when viewed from the face). These two views differ so markedly, we might have difficulty (given our limited perspective) believing that both are manifestations of the same physical object. Remember, we can't exit our flat world and see the quarter from all angles; we're stuck with the view we have.

There's only one way we would know that the edge-view and the face-view were of the same object, and that would be if we caught a glimpse of a rotating quarter. In that instance, we could observe the two forms becoming interchangeable. Remember, we have no intuitive feel for the third dimension, so the rotating quarter would still seem very strange, as if one object (a rectangle) were changing into another object (a circle) and back again before our very eyes. Eventually, some brilliant 2-D theorist postulates "the third dimension" and proposes that two entities are simply two different views of a single object that exists in a 3-D space beyond our mental comprehension. The transformation of rectangle to circle isn't a transformation at all, our flat Einstein argues, but merely a rotation of a single object in a space we can't personally touch or comprehend. The magical transformation has been reduced to geometry, albeit a geometry beyond our intuitive experience.

Particle physicists proposed a non-intuitive space to explain the identity of the proton and neutron, which they called isospin space. In isospin space, the proton and the neutron are two views of the same particle, called a nucleon. Depending upon how the nucleon is oriented in isospin space relative to our world, we will see it as either a proton or a neutron. Of course, the real test is to catch the nucleon in the act of spinning—changing from neutron to proton or back again. Indeed, we do see the nucleon "spin" during beta decay, when a neutron emits an electron and a neutrino to become a proton. The nucleon uses kinetic energy to change its isospin position, just like the spinning quarter. Pioneers like Gell-Mann greatly extended this concept by creating more elaborate abstract spaces and allowing particles to perform much more complex rotations that have no analogue in our physical world. We sit trapped in our four-dimensional world and let a handful of particles twist and turn into their myriad forms, like a few shards of colored glass spun into complex patterns by a kaleidoscope. What we call physics is really just an exercise in geometry. Eventually, physicists hope to reduce their domain to a single particle twirling in an enormously complex space. The search for the TOE is really the search for that space.

What does this have to do with networks? As it turns out, we can reduce network behavior to an exercise in unseen geometry, too. We can define an abstract space in which the coordinates are the connection weights. In this "weight space," the number of dimensions equals the number of possible weights; the weight space of a human brain may have ten trillion dimensions (yes, that's entirely possible).

One point in weight space would then represent the state of network knowledge at any given time. The network's knowledge resides in the values of its weights, and the weights define the position of the network in weight space. For example, if our network had only three connections, the weight space would have only three dimensions, and a specific set of weights (example: 0.2, 8.0, 1.5) would define both the current state of network learning and a single point in that three-dimensional weight space. Since learning requires changing weight values, it represents movement in weight space. Thus, all networks are on a journey through weight space in search of more perfect knowledge.

Many people criticized my juxtaposition of religion and evolution in my last two books, as if the two were mutually exclusive. They are not. Personally, I do believe in God, but not the God of Genesis.

Define an infinite weight space. In that space there must be one point at which a network knows all that it could ever possibly know, the single attractor representing perfect knowledge. That is where He must reside. Perhaps humans are closest to God, not because we are the smartest creatures per se, but because our brains, unlike any other network on the planet, have the potential for virtually infinite degrees of connectivity. Only humans can begin to understand the meaning of an endless weight space—and we sense His presence there too. Perhaps everything I know *is* wrong. Maybe brains are a little special after all.

People have many gods and many beliefs, but we are all on a common intellectual pilgrimage through a shared space, all attracted to the same sacred spot. As the eons pass, life forged its way through weight space buffeted by the Brownian noise that enveloped it. God did not create the world as we know it by a single edict. Instead, he gave us a set of glass beads, shook us like a giant pinball machine, and told us to play, and play we did. Networks grew from a few RNA enzymes to cells to brains to global networks, and even now, we reach across space and attempt to contact other worlds. The need to socialize doesn't stop at the stratosphere, and before our own Daisyworld collapses into its twilight inferno, we may yet reach our goal and come to rest in that final, perfect attractor.

Isn't that where life has been headed all along?

ADDENDA

In anticipation of future criticism of this book by, among others, experts in the field of network theory, let me reiterate what I've said earlier: I've had to simplify many concepts in order to make certain ideas accessible to the broadest possible range of readers. These simplifications don't invalidate the arguments presented here in the least, but I'll admit that they may cause some heartburn in AI experts.

For example, I could easily be criticized for treating the topic of "noise" with insufficient rigor, in that I don't always distinguish between neuronal noise (random errors in the activation states of neurons) and weight noise (errors in the transmission of neuronal data through weighted channels). In the first approximation, neuronal noise and weight noise appear similar. I recall an old "Saturday Night Live" skit involving a man standing on a New York subway platform who strains to decipher the incoherent messages blaring from a loudspeaker. The messages are hopelessly garbled, as anyone who has ever been on a subway platform already knows.

The scene then shifts to the control booth, where the two men who read the messages are conversing over lunch. When they speak, we realize that their real voices sound every bit as garbled as they did going over the public address system (uproarious laughter ensues). In this case, the "noise" wasn't in the communication channel (the PA system), but originated from the mealy-mouthed speakers themselves. Normally, we assume that the auditory scrambling of messages over a PA system derives from connection noise (the PA system itself), not from neuronal noise (the speakers), but the end result in both cases is identical: the receiving "neuron" on the subway platform hears the same noisy information.

Technically, activation noise and weight noise *aren't* the same thing. The "shaking" process known as simulated annealing (described in chapter 9) uses neuronal activation noise; in computer models, the McCulloch-Pitts neuronal noise level can be raised by increasing the slope of its sigmoidal activation curve, a technique that "blurs" the neuron's transition from on to off. The mathematical shape of the sigmoidal curve depends on a single variable, T, which becomes analogous to the temperature of physical systems; the simple step-function, wherein the transition from on to off is abrupt, is merely a sigmoidal curve where $T = 0$, a state with no thermal noise at all. As T increases, the slope in the sigmoidal curve becomes less steep and the neuronal noise rises. Some living networks use this same trick to shake themselves out of bad attractor basins. Years ago, I pointed out that various cytokines (such as interferons and interleukins) change the sigmoidal slope of the lymphocyte's antigen-response curve, effectively raising and lowering the "temperature" of lymphoid networks.

During genetic evolution, noise (mutations) occurs in the connection weights because our genes control the *connectedness* of enzymatic networks. On the other hand, non-genetic systems, like the brain (and the pre-genetic networks of the primordial earth), contain noise in both their neurons *and* their connections. In earlier chapters, I referred to *thought* as the spontaneous buffeting of connection weights by noise, but this is only approximately correct. In fact, *there must be two types of noise-driven thought: thought driven by activation noise and thought driven by weight noise.* A network uses neuronal noise to shake itself from one attractor to another; this is the annealing process. Annealing presumes that attractor states already exist, and so this form of thought can be more accurately described as decision-making. My brain already knows three different routes that I might travel from my office to my home, and the act of choosing the best route on a given day involves shaking myself into one of three established attractors, not creating a new attractor *de novo*. This choice can be accomplished through annealing.

Creating entirely new attractors using weight noise is a different matter. Normally, this is accomplished through external training (learning), but we now know that new attractors can also arise in intelligent networks in the absence of external input; we call this creativity. When Einstein induced the laws of General Relativity, his brain altered its own landscape to make attractors that had never existed before, and it did so without external training *per se*. These new attractor states could have arisen only via the sculpting of his brain's connection weight noise. Thus, we can formally define two types of thinking: (1) simple decision-making—the use of neuronal noise to shift a network among preexisting landscape basins, and (2) creative thought—the use of weight noise to sculpt entirely new landscape basins in the absence of external training. (This dichotomy loosely resembles the distinction between deductive and inductive reasoning.) When a bacterium "chooses" to activate one of its existing lactamase genes in the presence of penicillin, it's making a decision; but when an entire species of bacteria evolves a new lactamase out of whole cloth, it's using creative thought generated at the genome (connection weight) level. In both instances, the process depends upon a controlled exploitation of noise.

A complete discussion of noise would fill a large volume. Although AI experts and neuroscientists agree that noise is an integral part of intelligent networks (networks not only tolerate noise but mine it for their own purposes), mainstream evolutionary biologists still view mutational noise as a nuisance—one that just happens to give rise to evolution through dumb luck alone. Of course, evolution isn't dumb, but to see why this is so, we must understand the meaning of random noise in a network paradigm.

A second subject that I have glossed over somewhat (and that would fill yet another complete volume) is the role of *time* in network behavior. As we have seen, time enters the "network equation" in a number of ways: (1) in the form of time-dependent periodic/chaotic attractors; (2) in the learning rate (the velocity of weight changes per unit time); and (3) in the way in which activation states and weight changes are "updated," or installed into the system (synchronously, asynchronously, chaotically, or continuously). The student network was updated asynchronously, in that each student glanced at his or her meter every minute and altered the switches on that student's own personal schedule. Some network architectures can learn only at a certain rate before collapsing into instability. Similarly, some networks must be "updated" asynchronously, while others can be updated in other ways. These details, although important in a more comprehensive model of living intelligence, don't concern us here.

Ironically, the whole debate over whether the biosphere does or does not have an Intelligent Designer boils down to a question of relative network speed. Brains are very, very fast and ecosystem networks are very, very slow, yet they work in much the same way. On paper, brains and ecosystems use a similar architecture, but in the real world, we have trouble equating the intelligence of a brain with that of the evolving biosphere. We don't invoke Divine Intervention to explain the existence of electric toothbrushes, because we understand how a fast brain network can design and make such things without heavenly assistance. On the other hand, the existence of living organisms puzzles us, even though a network designed them, too. Yes, it was a slow network, but it had an incomprehensibly long time in which to think. To see the truth about evolution, we must be willing to abandon our intuition and accept that very slow networks can still accomplish very smart things, given a few hundred million years. I have trouble believing that slowly moving water could have carved the Grand Canyon, but it did. Still, to those who refuse to believe in erosion, the Grand Canyon must be the work of a god. Remember: different networks have different "clock speeds," but slow networks aren't necessarily dumber than fast ones.

Network purists will also point out that the "landscape" approach doesn't necessarily apply to all networks. That's quite true, but I believe that this approach applies to all *living* networks, and that's all I'm concerned about here. In this book, I've stressed the Hopfield approach to networks because Hopfield first showed how energy landscapes could be used as computational devices; life is not a Hopfield network in a literal sense. The Hopfield architecture belongs to the realm of electrical engineering, not to the realm of biology. Nevertheless, Hopfield's view of computation (as a gradient descent into an energy minimum) *does* belong in the realm of biology. The "genesis transition" from inorganic spin glass to a living spin glass seems plausible and seamless. The malleability of organic macromolecules endowed living spin glasses with the trainable connections they needed to imprint vast numbers of environmental patterns. The chemistry of carbon nicely facilitates the emergence of some very sophisticated spin glasses, but the physics remains the same in both the inorganic and the organic worlds.

I also need to clarify my use of the phrase "digital computer." Theoretically, a digital computer is any computing device that relies solely upon binary (step function) neurons. In the strictest sense, our student Hopfield device is still a digital machine because it consists of desks possessing only two states, on and off. In its original incarnation, the McCulloch-Pitts neuron has only two activation states, 0 or 1, and all machines made of such neurons are, by definition, digital machines. Earlier, I implied that the digital machines are equivalent to von Neumann machines (like the personal computer). While true in a practical sense, this equivalence is not true in any deeper sense.

The sigmoidal neuron, on the other hand, is an *analog* device. Because of its gentler transition between the on and off states, this neuron can assume any activation value between 0 and 1. The hallmark of an analog device is its ability to represent a continuous range of data points; a warning light on an automobile's dashboard is a digital device (on or off), but the engine temperature gauge is an analog device (presenting a continuous range of possible temperatures). Not to worry. Hopfield showed that his networks worked both for analog (sigmoidal) as well as digital (step function) neurons. The landscape paradigm applies to both digital and analog networks.

In the real world, particularly the biological world, sigmoidal (analog) neurons are the rule. As I've already asserted, sigmoidal neurons allow biological systems the flexibility to think more creatively. Sigmoidal neurons, because they still have two stable states, also retain the digital aspects of the McCulloch-Pitts neurons, while adding the additional power of analog computing. But the distinction between digital and analog networks may be even more profound.

Hava Siegelmann recently argued that digital networks, although they differ markedly from von Neumann computers in many aspects, are still Turing machines and that even the idealized Turing machine has limits. It can't compute everything. In contrast, Siegelmann proved that analog networks—more specifically, *recurrent analog networks*—can transcend the limitations of Turing's digital paradigm. In other words, recurrent analog networks not only do things differently, they can do things *better* when compared to digital machines. (For the record, recurrent networks are ones that allow any pattern of communication among its neurons. "Redcoat" networks, like the perceptron, aren't recurrent. Living systems, on the other hand, are.) Thus, life isn't merely a supercharged Turing machine, but may instead be something superior to a Turing machine altogether.

Siegelmann contends that analog networks that rely on statistical communication among their neurons (including the cytoplasm, the immune system, insect colonies, and Gaian ecosystems), although still superior to purely digital Turing machines, turn out to be inferior to those analog networks that use only deterministic connections (for example, the brain). Thus, we see another reason why brains may be special after all. By eliminating the element of chance from its neuronal connections, the nervous system becomes even closer to the "ideal" network. And, if we are to believe Siegelmann, the ideal network is the most powerful computational device that can ever exist. Is the human brain the closest thing to a perfect computational engine? At the beginning of this book, I questioned the supremacy of the human brain and here I am, at the book's very end, putting the brain back upon its pedestal. Well, at least now I have some theoretical rationale for doing so.

I readily acknowledge that the ideas presented here aren't fully formed. As an illustration, I can think of one glaring hole in the network model of the cell as I've described it: exactly how are enzymatic reactions "connected" inside a cytoplasm? In the third chapter, I implied that two reactions are connected if they share a common reactant and/or a common product; but if this is true, the connection weight would be both a function of the concentration of that shared reactant/product and a function of the enzyme kinetics. How would we model this mathematically? Who knows? I don't pretend to know all the answers, or even a lot of them. We can't look to Hjelmfelt's inorganic model for guidance here, since his "molecular computer" uses little pipelines to connect reaction chambers and doesn't work in the same fashion as a cytoplasm; the Hjelmfelt network is just a chemical version of a hardwired computer network operating on a molecular scale. (I should note that Hjelmfelt did publish a model of a connected cytoplasmic network based on enzymatic reactions, but it's too complex to discuss here.)

I still believe that the connection weights among enzymatic reactions inside a cell are written in the genome, but the exact nature of the chemical connections inside a cytoplasmic network remains a topic for future research. (And I haven't even mentioned the growing discipline of gene networks, which deals with a cellular subnetwork regulating the interactions of genes within the nucleus—that's another subject for an entire volume, perhaps two, but not for general readers.)

A full comprehension of life's many minds requires a deep, multidisciplinary knowledge beyond the intellectual scope of most living scientists (and certainly beyond the scope of my own inadequate mind). A "unified field theory" of living networks must necessarily span the fields of computer science, network theory, computational theory, AI, genetics, molecular and cellular biology, immunology, paleontology, evolutionary biology, neuroscience, psychiatry, and ecology, to name just a few. As such, any attempt by a single author to unite these fields into a common paradigm will be prone to errors and misinterpretations, particularly if he or she is writing for a general audience, and I fear that this book will prove to be no exception. As such, I will now hunker down and await the inevitable barrage of criticism by experts in other fields who will view me as they tend to view any outsider: as an amateur who dares walk upon their soil. To those experts, let me apologize in advance. I mean no harm. I would hasten to point out, though, that until such time as a separate discipline appears that is devoted solely to the study of *all* biological intelligence, from enzyme to ecosystem (and not limited to the intelligence of brains and brain-like devices), the study of the living process will remain dominated by amateurs. Any person who tries to bridge ten different fields is going look like an amateur in at least nine of them. Nevertheless, many good ideas have come from the minds of amateurs (including Darwin himself!), particularly if they aren't afraid to look foolish. (In my case, if I feared looking foolish, I wouldn't leave the house.)

There will be those experts who come armed to the teeth with detailed critiques ("Miyashita's monkey experiments didn't really *prove* the existence of attractor memory...a Hopfield network doesn't work in *quite* this way...asynchronous updating is more important than the author realizes...the fly has *sixteen* antibacterial proteins, not fifteen...his history of computing gets it all wrong"), to whom I will respond by asking: are the basic principles I'm stating correct, even if a detail or two (or three) may be wrong? If so, hold your tongues. One can criticize my particular interpretation of Miyashita's data, for example, but the basic prin-

ciple—our brains use attractor-based memory—can no longer be seriously questioned.

There are many deep problems in computational theory that I haven't even addressed. For example, are networks and serial machines really all that different? Theoretically, it can be argued that networks are also Turing machines. As such, network and serial computations may be just different versions of a single (but elusive) unified theory of intelligence. After all, rule-based PCs can run simulations of neural networks and the brain (a neural network) can write rule-based algorithms, so the Yang and Um boundary between the two paradigms is, once again, indistinct. Fortunately, these highly technical issues don't influence the arguments advanced in this book. The astronomer calculating the position of the planets cares only about *practical* gravitational theory; he doesn't care if gravitation and electromagnetism are linked in some highly abstract "superstring" model. Similarly, I'm interested only in *practical* network models, ones that can provide some immediate insight into the nature and magnitude of biological reasoning in creatures ranging from single cells to continental ecosystems. For now, I'm less concerned about highly speculative theories of computation. Computational theory may someday answer some very deep questions, such as precisely when and how the phase transition from inanimate network to living network took place, or why some brains appear conscious and others do not. Still, we must walk before we can run, and this book represents but a few halting steps in our journey to the ultimate understanding of intelligent life on earth.

I've tended to ignore thermodynamic topics, such as the fact that living systems appear to reduce their entropy in violation of thermodynamic dogma, because the laws of thermodynamics can be legitimately violated *if* we assume that life is intelligent from the beginning. We have no trouble understanding how the entropy of iron ore is reduced as it's converted into an automobile; humans, because of our intelligence, can manipulate thermodynamics to our heart's content. But many great minds have struggled to understand the paradoxical thermodynamics of the single cell, largely because they universally assume that the cell is stupid. If we assume the cell to be smart, then we don't have to worry about any perceived "violations" of the sacred laws of thermodynamics by living systems.

The topics covered here have been numerous and difficult. Because of the cross-disciplinary nature of this book, most readers will have encountered several chapters that proved "tough sledding," so to speak. Again, I apologize. I labored mightily for many long nights trying to boil tough ideas down to simple prose in order to make my ideas appealing to the largest possible audience while remaining faithful to the accuracy of the underlying science, yet I know that I haven't entirely succeeded on all counts. But to paraphrase Einstein once more, things can be made only so simple before they start to lose meaning. When faced with a choice between simplicity and meaning, I had to choose meaning. I began this book with a quote from Nietzsche's *Thus Spake Zarathustra,* a work he subtitled "For all and none." In a sense, my book is also for everyone and for no one at the same time. The subjects—life and intelligence—concern everyone, but few people possess the requisite knowledge to grasp the whole thing at one sitting.

Like any product of a living network, this work will contain errors—life is full of noise. The book is not intended to be a complete blueprint for living systems, nor do I advance a fully gestated model, exact and correct in every nuance and detail. If all I do is spur some interest and ignite a vigorous (or angry) debate, then

I'll have succeeded, even if many of my ideas turn out to be sorely mistaken. (At the very least, the reader will look at a bladder infection with new respect.) Currently, there is only a limited cross-disciplinary interest in a unified model of living networks—not nearly enough when we consider the profound scientific and philosophical implications of the two subjects involved, namely, life and intelligence.

Which, after all, still remain one and the same thing.

BIBLIOGRAPHY

Albert, Reka; Jeong, Hawoong; and Barabasi, Albert-Laszlo. "Error and attack tolerance of complex networks." *Nature* 406:378–81, 2000.

Allman, William. *Apprentices of Wonder: Inside the Neural Network Revolution.* New York: Bantam Books, 1989.

Arbib, Michael, editor. *The Handbook of Brain Theory and Neural Networks.* Cambridge, Massachusetts: MIT Press, 1998.

Bains, S. "A subtler silicon cell for neural networks." *Science* 277:1935, 1997.

Baker, G. L., and Gollub, J. P. *Chaotic Dynamics: An Introduction.* New York: Cambridge University Press, 1990.

Bliss, T. V. P. "Young receptors make smart mice." *Nature* 401:25–27, 1999.

Bliss, T. V. P., and Collinridge, G. L. "A synaptic model of memory: long-term potentiation in the hippocampus." *Nature* 361:31–39, 1993.

Bloom, Howard. *Global Brain.* New York: John Wiley & Sons, 2000.

Bonabeau, Eric; Dorigo, Marco; and Theraulaz, Guy. *Swarm Intelligence.* New York: Oxford University Press, 1999.

Bonabeau, Eric; Dorigo, Marco; and Theraulaz, Guy. "Inspiration for optimization from social insect behavior." *Nature* 406:39–42, 2000.

Bray, D. "Intracellular signaling as a parallel distributed process." *Journal of Theoretical Biology* 143:215–31, 1990.

Bray, Dennis. "Protein molecules as computational elements in living cells." *Nature* 376:307–12, 1995.

Bridges, B. "Hypermutation under stress." *Nature* 387:557–58, 1997.

Cairns, J.; Overbaugh, J.; and Miller, S. "The origin of mutants." *Nature* 335:142–45, 1988.

Capra, Fritoj. *The Web of Life.* New York: Anchor Books, 1996.

Carpenter, Gail. "Neural network models for pattern recognition and associative memory." *Neural Networks* 2:243–57, 1989.

Choe, Jae, and Crespi, Bernard, editors. *The Evolution of Social Behavior in Insects and Arachnids.* New York: Cambridge University Press, 1997.

Churchland, Patricia, and Sejnowski, Terrence. *The Computational Brain.* Cambridge, Massachusetts: MIT Press, 1992.

Crameri, A., et al. "DNA shuffling of a family of genes from diverse species accelerates directed evolution." *Nature* 391:288–91, 1998.

Dan, Y., and Mu-ming, P. "Hebbian depression of isolated neuromuscular synapses in vitro." *Science* 256:1570–73, 1992.

Darwin, Charles. *The Origin of Species.* Penguin reprint, London, 1859.

Dasgupta, D., editor. *Artificial Immune Systems and Their Applications.* New York: Springer-Verlag, 1998.

Davies, Paul, editor. *The New Physics.* New York: Cambridge University Press, 1989. (In particular, the article "Physics of far-from-equilibrium systems and self-organization" by Anthony Leggett, pp. 268–88.)

Dawkins, Richard. *The Selfish Gene.* New York: Oxford University Press, 1976.

Degn, H.; Holden, A.; and Olsen, L. F., editors. *Chaos in Biological Systems.* New York: Plenum Press, 1986.

Dembski, William A. *Intelligent Design.* Downers Grove, Illinois: Inter-Varsity Press, 1999.

Edis, Taner. "Darwin in mind." *Skeptical Inquirer* 25:35–39, 2001.

Epstein, Irving. "The consequences of imperfect mixing in autocatalytic chemical and biological systems." *Nature* 374:321–27, 1995.

Freedman, D., et al. "Categorical representation of visual stimuli in the primate prefrontal cortex." *Science* 291:312–16, 2001.

Freeman, W. J. "A proposed name for aperiodic brain activity: stochastic chaos." *Neural Networks* 13:11–13, 2000.

Gell-Mann, Murray. *The Quark and the Jaguar: Adventures in the Simple and the Complex.* New York: W. H. Freeman, 1994.

Gleick, James. *Chaos: Making of a New Science.* New York: Viking Penguin, 1987.

Goodman, L., and Gilman, A., editors. *The Pharmacological Basis of Therapeutics.* New York: Macmillan, 1975.

Gordon, Deborah. *Ants at Work.* New York: The Free Press, 1999.

Gordon, Deborah; Goodwin, Brian; and Trainor, E. H. "A parallel distributed model of the behavior of ant colonies." *Journal of Theoretical Biology* 156:293–307, 1992.

Gould, Stephen J., and Eldredge, Niles. "Punctuated equilibria: the tempo and mode of evolution reconsidered." *Paleobiology* 3:115–51, 1977.

Gregory, Richard L., editor. *The Oxford Companion to the Mind.* New York: Oxford University Press, 1987.

Griffiths, G. M.; Berek, C.; Kaartinen, M.; and Milstein, C. "Somatic mutation and the maturation of antibody response to 2-phenyl oxalzalone." *Nature* 312:271–75, 1984.

Grossberg, Stephen. "Nonlinear neural networks: principles, mechanisms and architectures." *Neural Networks* 1:17–61, 1988.

Hasselmo, M. E. "Runaway synaptic modification in models of cortex: implications for Alzheimer's disease." *Neural Networks* 7:13–40, 1994.

Hebb, Donald O. *The Organization of Behavior.* New York: John Wiley & Sons, 1949.

Hesse, Hermann. *The Glass Bead Game (Magister Ludi).* New York: Henry Holt, 1990.

Hjelmfelt, Allen; Schneider, F.; and Ross, John. "Pattern recognition in coupled chemical systems." *Nature* 260:335–38, 1993.

Hjelmfelt, Allen; Weinberger, Edward; and Ross, John. "Chemical implementation of neural networks and Turing Machines." *Proceedings of the National Academy of Sciences (USA)* 88:10,983–87, 1991.

Hoffman, G. W. "A neural network model based on the analogy with the immune system." *Journal of Theoretical Biology* 122:33–67, 1986.

Holland, J. H. *Adaptation in Natural and Artificial Systems.* Ann Arbor: University of Michigan Press, 1975.

Hölldobler, Bert, and Wilson, Edward O. *The Ants.* Cambridge, Massachusetts: Belknap Press, 1990.

Hopfield, John. "Neural networks and physical systems with emergent collective computational abilities." *Proceedings of the National Academy of Sciences (USA)* 79:2554–58, 1982.

Hopfield, John. "Neurons with a graded response have collective computational properties like those of two-state neurons." *Proceedings of the National Academy of Sciences (USA)* 81:3088–92, 1984.

Jackson, Robert C. "The kinetic properties of switch antimetabolites." *Journal of the National Cancer Institute* 85:538–45, 1993.

Jerne, N. K. "Toward a network theory of the immune system." *Ann. Immunol. Inst. Pasteur* 125C:373–89, 1974.

Kaiser, Dale. "Bacteria also vote." *Science* 272:1598–99, 1996.

Kandel, Eric R., and Schwartz, James H. *Principles of Neural Science.* New York: Elsevier/North-Holland, 1981.

Kauffman, Stuart. *At Home in the Universe.* New York: Oxford University Press, 1995.

Kohno, Y.; Berkower, I.; Minna, J.; and Berzofsky, J. "Idiotypes of anti-myoglobin antibodies: shared idiotypes among monoclonal antibodies to distinct determinants of sperm whale myoglobin." *Journal of Immunology* 128:1742–48, 1982.

Kohonen, T. "Self-organization and associative memory." Berlin: Springer-Verlag, 1984.

Kuhn, Thomas S. *The Structure of Scientific Revolutions.* University of Chicago Press, 1996.

Kurzweil, Ray. *The Age of Spiritual Machines.* New York: Viking, 1999.

Laubach, M., et al. "Cortical ensemble activity increasingly predicts behavior outcomes during learning a motor task." *Nature* 405:567–70, 2000.

Levin, B., et al. "The population genetics of antibiotic resistance." *Clinical Infectious Diseases* 24 (Supplement 1): S9–S16, 1997.

Levy, Pierre. *Collective Intelligence.* New York: Plenum Trade, 1997.

Lewin, Roger. *Complexity.* University of Chicago Press, 1999.

Lisberger, S., and Sejnowski, T. "Motor learning in a recurrent network model based on the vestibulo-ocular reflex." *Nature* 360:159–61, 1992.

Lisboa, P. *Neural Networks.* New York: Chapman & Hall, 1992.

Liu, Y.; Bona, C.; and Schulman, J. "Idiotypy of clonal responses to influenza virus hemaglutinin." *Journal of Experimental Medicine* 154:1525–38, 1981.

Lovelock, James. *The Ages of Gaia.* New York: W. W. Norton & Co., 1988.

Mandelbrot, Benoit. *The Fractal Nature of Geometry.* New York: W. H. Freeman, 1982.

Manschreck, T. C.; Maher, B. A.; Milavertz, J. L.; Ames, D; Weisstein, C.; and Schneyer, M. L. "Semantic priming in thought-disordered schizophrenic patients." *Schizophrenia Research* 1:61–66, 1988.

Marijuan, Pedro. "Enzymes and theoretical biology: sketch of an informational perspective of the cell." *Biosystems* 25:259–73, 1991.

Markarenkov, V., and Clinas, R. "Experimentally-determined chaotic phase synchronization in a neuronal system." *Proceedings of the National Academy of Sciences (USA)* 95:15,747–42, 1998.

McAdams, Harley, and Shapiro, Lucy. Circuit simulation of genetic networks. *Science* 269:650–56, 1995.

McClelland, James L.; McCaughton, B. L.; and O'Reilly, R. C. "Why there are complementary learning systems in the hippocampus and neocortex: insights from the success and failures of connectionist models of learning and memory." *Psychological Review* 102:419–57, 1995.

McCulloch, W. S. and Pitts, W. "A logical calculus for the ideas immanent in nervous activity." *Bulletin of Mathematical Biophysics* 5:115–33, 1943.

Mel, S., and Mekalanos, J. "Modulation of horizontal gene transfer in pathogenic bacteria by in vivo signals." *Cell* 87:795–98, 1996.

Metzger, D.; Miller, A.; and Sercarz, E. "Sharing of idiotypic marker by monoclonal antibodies specific for distinct regions of hen lysozyme." *Nature* 287:540–42, 1987.

Minsky, M., and Papert, S. *Perceptrons.* Cambridge, Massachusetts: MIT Press, 1969.

Miyashita, Y. "Neuronal correlate of visual associative long-term memory in primate temporal cortex." *Nature* 335:817–20, 1988.

Mjolsness, Eric; Sharp, David; and Reinitz, John. "A connectionist model of development." *Journal of Theoretical Biology* 152:429–53, 1991.

More, M., et al. "Enzymatic synthesis of a quorum-sensing autoinducer through the use of defined substrates." *Science* 272:1655–58, 1996.

Moxon, E., and Thaler, D. "The tinkerer's evolving tool-box." *Nature* 387:659–62, 1997.

Murray, P.; Swain, S.; and Kagnoff, M. "Regulation of the IgM and IgG anti-dextran B1255S response: synergy between IFN, BCGF and IL-2." *Journal of Immunology* 135:4015–20, 1985.

Myers, M. P., et al. "Light-induced degradation of TIMELESS and entrainment of the Drosophila circadian clock." *Science* 271:1736–40, 1996.

Oudin, J., and Cazenave, P. "Similar idiotypic specificities in immunoglobulin fractions with different antibody functions or even without detectable antibody function." *Proceedings of the National Academy of Sciences (USA)* 68:2616–20, 1971.

Parberry, Ian. *Circuit Complexity and Neural Networks.* Cambridge, Massachusetts: MIT Press, 1994.

Penrose, Roger. *The Emperor's New Mind.* New York: Oxford University Press, 1989.

Perelson, A. "Immune network theory." *Immunological Reviews* 110:5–36, 1989.

Pinker, Steven. *How the Mind Works.* New York: W. W. Norton & Co., 1997.

Pritchard, L., and Dufton, M. "Do proteins learn to evolve? The Hopfield network as a basis for the understanding of protein evolution." *Journal of Theoretical Biology* 202:77–86, 2000.

Rabinowitz, M. I., and Abarbanel, H. D. "The role of chaos in neural systems." *Neuroscience* 87:5–14, 1998.

Raff, M. "Immunologic networks." *Nature* 265:205–207, 1977.

Rajewsky, K.; Schmirrmacher, V.; Nasi, S.; and Jerne, N. "The requirement of more than one antigenic determinant for immunogenicity." *Journal of Experimental Medicine* 129:1131–38, 1969.

Reich, J. G., and Sel'kov, E. E. "Mathematical analysis of metabolic networks." *FEBS Letters* 40:S119–27, 1974.

Resnick, Mitchel. *Turtles, Termites and Traffic Jams.* Cambridge, Massachusetts: MIT Press, 2000.

Roitt, I.; Brostoff, J.; and Male, D., editors. *Immunology*. Philadelphia: Mosby, 1998.

Rosenblatt, Frank. "The perceptron: a probabilistic model for information storage and organization in the brain." *Psychological Review* 65:386–408, 1958.

Rummelhart, David E.; Hinton, Geoffrey E.; and Williams, Ronald J. "Learning representations by back-propagating errors." *Nature* 323:533–36, 1986.

Rummelhart, David; McClelland, James; and the PDP Research Group. *Parallel Distributed Processing*, volume 1. Cambridge, Massachusetts: MIT Press, 1986.

Ryan, J. "Lesion of the subthalamic nucleus does not cause chaotic firing patterns in basal ganglia region of rats." *Brain Research* 873:263–67, 2000.

Schrödinger, Erwin. *What Is Life?* Cambridge, UK: Cambridge University Press, 1967.

Servan-Schreiber, D.; Printz, H.; and Cohen, J. D. "A network model of catecholamine effects: gain, signal-to-noise ratio and behavior." *Science* 249:892–95, 1990.

Sharma, Kamal; Leonard, Ann; Letteri, Karen; and Pfaff, Samuel. "Genetic and epigenetic mechanisms contribute to motor pathfinding." *Nature* 406:515–19, 2000.

Siegelmann, Hava. *Neural Networks and Analog Computation: Beyond the Turing Limit*. Boston: Birkhauser, 1999.

Smith, John Maynard, and Szathmary, Eors. *The Origins of Life*. New York: Oxford University Press, 1999.

Sniegowski, P.; Gerrish, P.; and Lenski, R. "Evolution of high mutation rates in experimental populations of E. Coli." *Nature* 387:703–705, 1997.

Sole, Ricard, and Marin, Jesus. "Macroevolutionary algorithms: a new optimization method on fitness landscapes." *IEE Transactions on evolutionary computations* 3, 1999.

Spitzer, Manfred. *The Mind Within the Net*. Cambridge, Massachusetts: MIT Press, 1992.

Stoop, R., et al. "When pyramidal neurons lock, when they respond chaotically, and when they like to synchronize." *Neuroscience Research* 36:81–91, 2000.

Taddel, F., et al. "Role of mutator alleles in adaptive evolution." *Nature* 387:700–702, 1997.

Tank, David W., and Hopfield, John. "Collective computation in neuronlike circuits." *Scientific American* 257:104–14, 1987.

Thorpe, S., and Fabre-Thorpe, M. "Seeking categories in the brain." *Science* 291:260–62, 2000.

Touitou, Yvan, editor. *Biological Clocks*. New York: Elsevier, 1998.

Turing, Alan. "On computable numbers with an application to the Entscheidungsproblem." *Proceedings of the London Mathematical Society* 2:230–65, 1936.

Turing, Alan. "Computing machinery and intelligence." *Mind* 59: 433–60, 1950.

Vertosick, Frank T. "The immune system as a neural network: an alternative approach to immune cognition." *SigBio Newsletter* 12:4–6, 1992.

Vertosick, Frank T. "Fluid networks as a model of intelligent biological systems," in Ray Paton, editor, *Computing with Biological Metaphors*. New York: Chapman and Hall, 1994, pp. 156–65.

Vertosick, Frank T., and Kelly, Robert H. "Immune network theory: a role for parallel distributed processing?" *Immunology* 66:1–7, 1989.

Vertosick, Frank T., and Kelly, Robert H. "The immune system as a neural network: a multiepitope approach." *Journal of Theoretical Biology* 150:225–37, 1991.

Von der Malsburg, C. "Self-organization of orientation-sensitive cells in striate cortex." *Kybernetik* 14:85–100, 1973.

Walsh, Christopher. "Molecular mechanisms that confer antibacterial drug resistance." *Nature* 406:775–81, 2000.

Waltrip, M. Mitchell. *Complexity*. New York: Simon & Schuster, 1992.

Watson, James D., et al., editors. *Molecular Biology of the Gene*. Menlo Park, California: Benjamin/Cummings, 1987.

Weinberg, Robert, editor. *Oncogenes and the molecular origins of cancer*. Cold Spring Harbor, New York: Cold Spring Harbor Laboratory Press, 1989.

Wilson, Edward O. *The Insect Societies*. Cambridge, Massachusetts: Belknap Press, 1971.

INDEX

absolute zero, 216

accelerated mutation, 46

acetylcholine, 242

action potential, 236–38, 253–54

activation rule, 178, 294

activation state, 178–79, 327

adaptation, 42, 48, 62, 65, 67–69

adaptive immunity, 69, 85–86

adenine, 27

affinity, 78, 184

alanine, 38

albedo, 281, 285

Allman, William, 161

Alzheimer's disease, 125, 233–34, 249–50, 257

amino acids, 27–31, 33, 58, 71, 207, 211, 241

amnesia, 233–34

ampicillin, 52

amplifiers, 102

analog networks, 330

annealing, 296, 325, 326

ant colonies, 247, 270–78, 290

antibiotics, 22–26, 34–37, 50, 68

antibodies, 69–80, 84, 86, 201, 315–16. *See also* immune system

antigens, 12, 69–70, 78, 79, 84, 184, 201, 218

aperiodic crystal, 153

artificial intelligence (AI), 140, 143–44, 151, 190, 281, 313, 324, 327

associative learning, 248, 260, 262, 265

attractor-based memory, 255–56

attractor basin, 111, 266, 285–86, 295, 304

attractor landscape, 299

attractor recall, 185

auto-antibodies, 74–75

autocatalysis, 100–102, 104–5

axons, 187–89, 237–39

background noise, 216–17, 265

back-propagation of errors (backprop), 192–94, 202–4, 214, 215

bacteria, 14, 15–16, 21–55
  adaptation of, 48, 62
  antibiotics and, 22–26, 34–37, 50, 68
  cancer cells versus, 129
  communities of, 48, 271
  connectivity of, 185
  intelligence of, 54–55, 250
  pathogenic, 49–52
bacteriophage, 40
bacteriophage transduction, 51
base sequence, 29
B cells, 81–82
Beckett, Samuel, 265
bees, 247, 272
Belousov-Zhabotinski (BZ) reaction,
  100, 104–5
beta-lactam ring, 38–39
binary systems, 98–100
binding sites, 70–72, 95–96, 101,
  184–85
bioluminescence, 43–44, 63
bi-stable devices, 98–107, 109, 140–41,
  152
Bloom, Howard, 288, 290
Boltzmann distribution, 96
Bonabeau, Eric, 275–77
brain, 16, 155, 231–69
  genetic learning, 250–51, 277–78,
    301, 302–4
  network theory and, 232–37,
    254–58
  neurons and, 236–39, 246, 250, 260
  neurotransmitters and, 241–44, 246,
    247, 253–54, 269
  noise and, 258–66
  non-genetic learning, 250–54, 257,
    268
  synapses, 238–41, 243–44, 247–49,
    251–52, 268
brain chauvinism, 5–8, 11–13, 14, 46,
  113, 119, 180, 206–7, 251, 269

Bray, Dennis, 121
Brownian motion, 216–17, 258
Brownian noise, 323
butterfly effect, 311
BZ (Belousov-Zhabotinski) reaction,
  100, 104–5

Cairns, John, 45–46
cancer, 53, 73, 81, 87–92, 103, 122–31,
  227, 263, 296–97, 316–18
carbon dioxide, 279
catalysts, 94–95
cells, 14
  chemistry of, 90–93, 99
  intelligence of, 12–13, 53
  socialization of, 62–64
  See also cancer
cellular differentiation, 198
cellular immunity, 81
cellular regulatory proteins, 30
central processing unit (CPU), 145–48
cerebral cortex, 69, 254, 256, 266, 268
C-genes, 73–74
chaos, 305, 308–14
chaotic attractors, 308–9, 311–12
chemical bonding, 95
chemical reactions, 90–104, 107–8,
  110–13
chlorpromazine, 317
Choe, Jae, 277
cholera, 64
chromosomes, 30, 117
cilia, 21–22
ciprofloxacin, 52
cloning, 116–18, 119–20
closed system, 92
codons, 29
Cohen, Jonathan, 262
coincidence detectors, 252
competition, 61, 63, 67–68, 213–14
computational speed, 189–90

computational theory, 149–50, 251, 264–65, 269, 333
connectionism, 140, 142, 143–47, 155, 157, 177–78, 180–81, 281, 287, 317–19. *See also* networks
connectionist devices. *See* neural networks (parallel processors)
connection weights, 155–60, 167–68, 173–74, 180, 184–87, 190–94, 197, 202–4, 209–12, 216, 217, 218–21, 243–48, 250, 266, 268–69, 273, 285, 294, 312–13, 322, 325–26, 330
connectivity, 138–39, 182–83, 185, 216
consciousness, 13, 60–61, 287–89
content-addressable memory, 255–56
content recall, 256
cooperation, 61, 90
Copernicus, 4
cortex, 69, 254, 256, 266, 268
cowpox, 59
creationism, 16, 174, 175, 206
creativity, 180, 263–65, 326
Crespi, Bernard, 277
Crick, Francis, 27
cross-inhibition, 101–2
cybernetics, 208–9
cytokines, 82, 316
cytoplasm, 97, 120, 121–22, 125, 130, 155, 174, 184, 188, 211, 237, 271
cytoplasmic chemical networks, 118–19
cytosine, 27

Daisyworld (computer model), 281–86, 298, 323
Darwin, Charles, 205, 214, 332
Darwinian selection, 16, 34 37, 53, 66, 67–69, 79, 86, 205, 213–14, 216, 217, 218, 219, 223, 224, 226–27, 253, 257, 258
Dawkins, Richard, 53, 205–6, 215, 217, 219–20, 224

dementia, 125, 233–34, 249–50, 257
dendrites, 187, 188, 237, 238–39
deoxyribose, 27
differentiation, 117–19, 121–22
digital computing, 142–43, 145–46
diphtheria, tetanus, and pertussis (DPT), 79, 316
DNA (deoxyribonucleic acid), 27–33, 42, 51, 54, 119, 205, 207–8, 211–13, 219, 220, 224, 258, 263, 294–95, 298
    errors in, 44, 45
    germline, 72 75, 120
    modification of, 67, 72–73, 80, 125–28
dopamine, 242–43, 260–63, 269
Dorigo, Marco, 275–77
double helix, 27, 34
dreaming, 266
dryware, 140
dualism, 135–76

E. coli, 36–38, 40, 45, 46, 48, 49–51, 55
edge-view, 320–21
Edis, Taner, 299
Einstein, Albert, 4, 216, 217–18, 319–20, 321, 326, 334
Eldredge, Niles, 225–27
electronic models of networks, 306–8
elementary particle theory, 41
embryogenesis, 117, 119–20, 122–23
emergent properties, 16–17, 42, 60–64, 137–39, 215, 281
Emerson, Alfred, 272
energy surface, 163
energy valley, 164
ENIAC, 146, 147–48
entropy, 95, 333–34
enzyme kinetics, 225
enzymes, 62, 94–131, 142, 184, 186, 188, 314

epidermal growth factor (EGF), 127–28
epithelial cells, 126–27
epitope, 82–83
equilibrium, 91–97, 103
equilibrium landscape, 110–16
ergonomic matrix, 273
ergonomic scaffolding, 275
erythroblastosis, 128
E surfaces, 111–16, 119, 121, 124–25,
    152–54, 163, 166–72, 185, 197, 200,
    209–11, 216, 217, 221, 226, 274, 300,
    301
ethical systems, 7–8
Euclid, 4
eusocial networks, 272, 277
evolutionism, 8, 16, 205–6, 219–20, 281,
    298–99, 301, 327
experience, 9
extinction, 66

face-view, 320–21
feedback control, 101
feedback loops, 101–2
fertilization, 116–17
firing rate, 236–37
Fleming, Alexander, 23, 25, 26
flip-flop circuits, 101–2
Florey, Howard, 23, 26
Ford, Henry, 212
fractals, 313–14
Freedman, David, 256–57

Gaia, 278–86, 290, 291, 304, 314, 318
gametes, 116
gamma-amino-butyric acid (GABA),
    253–54, 257
gastrointestinal systems, 64
Gell-Mann, Murray, 40–41, 42, 321
gene exchange, 39–40, 42, 44, 47, 48
generalization, 10
General Theory of Relativity, 319–20,
    326

genes, 28–29, 30, 53
genetic chauvinism, 206–7
genetic code, 8, 34, 71–81, 84–85, 206,
    207, 247, 301. See also DNA
    (deoxyribonucleic acid)
genetic learning, 250–51, 277–78, 301,
    302–4
genome, 327
genotype, 118–20
gentamicin, 50–51
germline DNA, 72–75, 120
Gleick, James, 305
global learning rule, 202–4, 217–19
global mind, 288–89
glutamatergic synapses, 252
Gödel's theorem, 299
Golding, William, 278
Goodwin, Brian, 275
Gordon, Deborah, 275, 277–78
Gould, Stephen Jay, 53, 225–27
granulocyte stimulating factor (GSF),
    130–31
gravity, 103, 112, 137
greenhouse effect, 279
group theory, 40–41
guanine, 27

hardwired networks, 181–82, 188–91,
    241, 244–45
heat, 94–97, 296
Hebb, D. O., 168–70
Hebbian learning, 168–71, 175, 185,
    199, 202–4, 209, 215, 217, 218, 222,
    223, 225, 252–54, 257, 268, 300
Heisenberg Uncertainty Principle,
    308–9, 313
hemoglobin, 28, 33
Hesse, Hermann, 292–93, 307
hidden neurons, 198–200, 201, 233, 235,
    248–49, 300–301
Hillis, David, 175–76
Hinton, Geoffrey, 175–76, 192

hippocampus, 266–68
HIV, 81–82, 126
Hjelmfelt, Allen, 100, 109, 331
Hölldobler, Bert, 273
homogeneous networks, 234–35
honey bees, 247, 272
Hopfield, John, 140, 148–49, 151–56,
    170, 303
Hopfield networks, 160–67, 175–76,
    177, 192, 197–98, 209, 210, 234–35,
    258, 284, 307, 328–29, 332–33
hormone receptors, 77
hormone signaling, 44
human genome, 120, 208, 245–47, 250,
    302
humoral immunity, 81
hydrogen, 91–94, 111, 279, 311
hypermutation, 44–48, 51, 263

ideal networks, 330
idiotype labeling, 315–16
idiotypes, 202
idiotypic interactions, 82
immune system, 12–13, 14, 22, 51–52,
    55, 56–86, 155, 185, 189–90,
    201–2, 217–19, 245–46, 257, 315,
    316–17
  adaptive immunity and, 69, 85–86
  defined, 58
  emergent behaviors and, 60–64
  evolution and, 67–69
  genetic code and, 71–81, 84–85
  infant, 75–76
  non-adaptive immunity and, 64–66,
    85–86
  pattern recognition and, 59, 79–80
  structural barriers and, 63–64
  vertebrate, 66–86
immunoglobulins, 81–83
influenza, 21–22, 72, 75
information theory, 96–97
input neurons, 198–200, 235

input-output (I/O) devices, 97–98,
    106–9, 112, 113, 129–30, 141–42,
    190, 201, 236, 273, 297–98
insects, 247–48, 314
  ant colonies, 247, 270–78, 290
  bees, 247, 272
instinct, 12, 85, 115, 244–49
insulin, 31, 32, 123
integrated circuits, 144, 148
intelligence, 9–13
  evolution of, 53–54
  limits of, 54–55
  machine, 13–15, 53–54
  nature of, 5
  operational definition of, 9
intelligent aggregates, 139
intercellular communication, 46, 47,
    48–49
Internet, 287–88, 290
invertebrates
  brains of, 248
  learning by, 302–3
iodate-arsenous reaction, 99–100,
    104–5, 107–8, 109
iron, 28
isospin space, 320, 321

Jackson, Robert, 100–101, 103, 131,
    316–17
Jenner, Edward, 58–59
Jerne, Niels, 201–2, 314

Kauffman, Stuart, 226
knowledge, exponential growth of, 41
Kohonen, Teuvo, 254–55
Kohonen network, 254–55
Kubrick, Stanley, 53–54
Kuhn, Thomas, 4
Kurzweil, Ray, 312–13

lactamases, 38–39, 40, 42, 48, 49, 53,
    326–27

lactone, 43
language, 14, 58, 212–13, 302
large-scale networks, 191
L-dopa, 242, 261
learning process, 61, 301
learning rate, 221–22, 299–300, 327
learning rules, 202–4, 217–19, 221,
    222–25, 294, 301–2
leukemia, 128
Levy, Pierre, 288
Lie unitary groups, 41
limestone, 280
liver cells, 76, 124
living networks, 17, 214–16
local competition, 213–14
local learning rule, 202–4, 217, 218, 221,
    222–25
long-term memory, 57–58, 233, 254,
    267–68
long-term potentiation (LTP), 252, 268
Lovelock, James, 278–86, 290–91
lung cells, 124–25
lux, 43–44
lymphocytes, 73–86, 142, 155, 184, 186,
    201–2, 218, 257, 274, 314–15

machine intelligence, 13–15, 53–54
macromolecules, 27–32, 55, 62–63, 184,
    207–9, 211, 221. See also DNA
    (deoxyribonucleic acid)
magnetic crystals, 152, 154, 161
malignant transformation, 89
Mandelbrot, Benoit, 313–14
mapping, 153
Markham, Beryl, 54
Maxwell, James Clerk, 95–97
Maxwell's demon, 96–97, 265
McClelland, James, 175–76, 266
McCulloch, Warren, 141, 307
McCulloch-Pitts neurons, 141–42, 148,
    152, 154–55, 157, 160, 162, 175,

178, 179, 200–201, 236, 239, 275,
    287, 325, 329–30
membrane receptors, 125
memorization, 10
memory cells, 57–59, 80
methicillin, 52
Meyer, Jean-Arcady, 272
Minsky, Marvin, 143–44, 148
missense mutation, 33
Miyashita, Yasushi, 256, 332–33
mobile networks, 181–82, 186
molecular intelligence, 105
molecular patterns, 12
mono-stable devices, 105–8
Morse code, 145
mRNA (messenger RNA), 29, 31, 54
multicellular organisms, 130, 271
multi-stable devices, 104, 107, 109
mutagens, 126
mutations, 32–37, 40, 43–49, 117,
    125–28, 213, 218, 263, 294
myoglobin, 315–16

neo-Darwinism, 205, 219
nervous system, 219, 220
network devices. See neural networks
    (parallel processors)
network learning, 213–22
networks
    basic components of, 178
    as basis of living intelligence, 17, 216
    concept of, 108–10
    defined, 17, 109
    digital computing versus, 142
    intelligent aggregates as, 139
    internal equilibrium landscapes of,
        112
    non-genetic, 291
    post-genetic, 220, 221–22, 249–50
    pre-genetic, 220, 221–22
    See also specific types of networks

network theory, 143–44, 160–76, 221, 226, 232–37, 254–58, 276
neural networks (parallel processors), 113, 121, 143, 146–47, 148–49, 152, 173, 178, 191–92, 196–204, 254–55, 333
neuronal noise, 324–26
neurons, 99, 109–10, 146–47, 155, 181, 190, 196–202, 206–13, 250, 274, 294
    alliances among, 224
    in Daisyworld model, 285
    definition of generic, 206
    forms of, 206
    hidden, 198–200, 201, 233, 235, 248–49, 300–301
    input, 198–200, 235
    McCulloch-Pitts, 141–42, 148, 152, 154–55, 157, 160, 162, 175, 178, 179, 200–201, 236, 239, 275, 287, 325, 329–30
    output, 198–200, 235
    parts of, 237–39
    as selfish, 217, 222–25
neurosurgery, 234
neurotransmitter receptors, 238–39, 243, 246, 252, 269
neurotransmitters, 241–44, 246, 247, 253–54, 269
neutrons, 320
nitric oxide, 123
N-methyl-D-aspartate (NMDA), 252–54, 262–63
noise, 16, 180, 258–66, 294, 295, 298, 299, 313, 316, 324–25, 334
non-adaptive immunity, 64–66, 85–86
non-genetic learning, 246–47, 250–54, 257, 268, 301–4
non-genetic signals, 42–43
nonlinear systems, 41–44, 47, 97–98
nonsense mutations, 33–34

norepinephrine, 242
nucleic acids, 27, 30, 31, 104, 105, 220
nucleon, 321
nucleotide bases, 27
nucleus, in cloning, 116–18, 119–20

oncogenes, 125–27
on/off rule, 178–79
open system, 93–94
oscillators, 307
Oudin-Cazenave phenomenon, 316
output error, 192–93
output neurons, 198–200, 235
ova, 12–13, 80, 116, 119
oxacillin, 50, 51, 52
oxygen, 28, 91–94, 111, 279

Papert, Seymour, 143–44, 148
paradigm shifts, 4
parallel-distributed processors. See neural networks (parallel processors)
parallel processors. See neural networks (parallel processors)
Parberry, Ian, 269
Parkinson's disease, 242, 249
party networks, 182–83, 186–91, 204, 211, 239–40, 274, 285
Pasteur, Louis, 4
pathogenic bacteria, 49–52
pathogen memory, 57–59
pattern completion, 115, 124, 172, 195–96
pattern recognition, 10–13, 52, 59, 79–80, 114–15, 130, 194
Pauling, Linus, 4
pendulum, 305–6, 308–14
penicillin, 23–24, 34, 37, 38, 39, 42, 49, 50, 52, 65, 326–27
Penrose, Roger, 299, 312
peptides, 241

peptidoglycan, 37–38
perceptrons, 143–44, 148, 150–51, 199, 247–48, 330
perceptual categorization, 256
periodic attractors, 305–6, 311–12
perpetual motion device, 96
phagocytes, 70, 80
phase transitions, 289
phencyclidine (angel dust), 262–63
phenotype, 118, 120–22, 131
phenotypic landscape, 118–19, 123–24
photons, 309
photosynthesis, 64
Pitts, Walter, 141, 307
plasmid, 50
pluripotential cells, 117, 122–23
point attractors, 305–7, 311–12
point mutation, 32–33, 47
polio, 64, 80–81
polymers, 27–30, 220
product, 92
programs, 142
proline, 33
protein enzymes, 104
protein hormones, 123
protein networks, 212–13
proteins, 27–30, 58, 71–72, 211
protons, 320
Ptolemy, 8
punctuated equilibrium, 225–27, 298

Q/K receptor, 252
quantum mechanics, 312
quasi-equilibrium, 101
quorum sensing, 43–44

random events, 215–16, 299
random mutation, 37, 39–40, 47, 218
random networks, 167
random surface, 163–64
reactants, 92

reason, 9, 42
reflexes, 9, 12, 262
regulatory sites, 101, 127
Reich, J., 101
relative stability, 153
relays, 145–46, 147
reproducibility, 167
resistance, to antibiotics, 24, 34–37, 50, 65
resistors, 161–67, 209–10
retroviruses, 126
reverse reaction, 93
reversible memory, 154
ribosomes, 29, 31, 33–34, 35–36, 54
RNA (ribonucleic acid), 29–31, 54, 104, 126, 323
Rosenblatt, Frank, 144
rule-based devices. See serial processors
Rummelhart, David, 175–76, 192

savants, 195–96
scale-free networks, 234
scale invariance, 314
schizophrenia, 262, 263, 265, 266, 296–98, 317
Schneirla, Theodore, 247
Schrödinger, Erwin, 153
scientific creationism, 174
Sejnowski, Terrence, 175–76
self-organization, 175, 203, 266
Sel'kov, E. E., 101
semantic priming, 260–61
semantic priming effect, 261, 262
semiconductors, 147
serial processors, 143–45, 146–48, 149–50, 155, 166, 173, 191–95, 196, 233, 263, 289–90, 333
serine, 29, 33
serotonin, 242
Servan-Schreiber, David, 262

Seurat, Georges, 264–65
sexual reproduction, 39–40, 46
short-term memory, 266–68
sickle cell anemia, 33
Siegelmann, Hava, 330–31
sigmoidal neurons, 329–30
sigmoidal rule, 179–80
signal-to-noise ratio, 258–66, 294–98, 316, 318
simulated annealing, 296, 325
sleep states, 265–66
smallpox, 59
small-scale networks, 191
smart networks, 151
social intelligence, 90
socialization, 60–64
soil bacteria, 39
Sole, Ricard, 186
solid-state crystals, 153
souls, 7
sperm, 80, 116
spin glasses, 152–53, 154, 183
Spitzer, Manfred, 120
squamous cell cancer, 129
staphylococcal cultures, 23, 37, 39, 40, 46, 48, 49–52
statistical probability, 186–87
steady-state equilibrium, 91–92
stem cells, 117
streptomycin, 35–36, 47, 50
stress
    hypermutation and, 46, 47
    of microbial infection, 68–69
structural barriers, 63–64
supercomputers, 24–25, 84
superorganisms, 270–91
    ant colonies, 247, 270–78, 290
    Daisyworld, 281–86, 298, 323
    Gaia, 278–86, 290, 291, 304, 314, 318
    global mind, 288–89, 290–91
    Internet, 287–88, 290

superstring model, 333
surface catalysis, 94–95, 100
swarm intelligence, 275–77
switching states, 157, 237
switch pharmaceuticals, 103
symbiosis, 122
synapses, 238–41, 243–44, 247–49, 251–52, 268

targeted hypermutation, 46
T-cell antigen receptors (TCARs), 69, 81–82
temperature, 94–97
T helper cells, 81–82
Theory of Everything (TOE), 293, 322
Theory of Relativity, 319–20, 326
Theraulaz, Guy, 275–77
thermodynamics, 95–97, 111, 333–34
thought, 216, 217
thymidine, 27
time, 16, 327–28
Trainor, Lynn, 275
transcription, 29–31, 34
transduction, 40, 46, 51
transformation, 40, 46, 51
transistors, 99, 144, 145, 147–48, 181, 237, 268–69
translation, 30, 34
triplet code, 29–31, 33–34
tuberculosis, 35–36, 47
Turing, Alan, 13–14, 140–41, 143, 288
Turing machines, 84, 140–41, 142–43, 312, 330, 333
Turing test, 13–14, 26, 55, 72, 140–41
2001: A Space Odyssey (movie), 53–54
typhoid, 64

unified field theory, 143, 221, 331–32

V. fischeri, 43–44, 60–61, 63–64
vaccines, 56–59, 68, 79

vacuum tubes, 146, 147, 166, 237
v-erbB, 128
vertebrates
  brain of, 248, 250
  immune systems of, 66–86
  learning by, 302–3
V-genes, 72–76, 78–81
violence, 63
viruses, 21–22, 125, 250
  computer, 32
  as go-betweens, 40
  impact on bacteria, 31–32, 185
  vaccines and, 56–59, 68
von Neumann, John, 143, 144–45, 146
von Neumann machines, 142–43, 195,
  234, 329, 330

V-regions, 70–72, 74, 75, 77, 78, 79,
  82–83, 201, 218

water, 91–94, 111, 279
Watson, Andrew, 281, 286
Watson, James, 27
weight matrix, 157, 159–60
weight noise, 324–25
weight space, 322–23
wetware, 140
Wheeler, William Morton, 271–72
Williams, Ronald, 192
Wilson, Edward O., 247, 272–73, 288,
  314
wisdom, 8–9

p/04